# Firefighting Toys

## 1940s-1990s

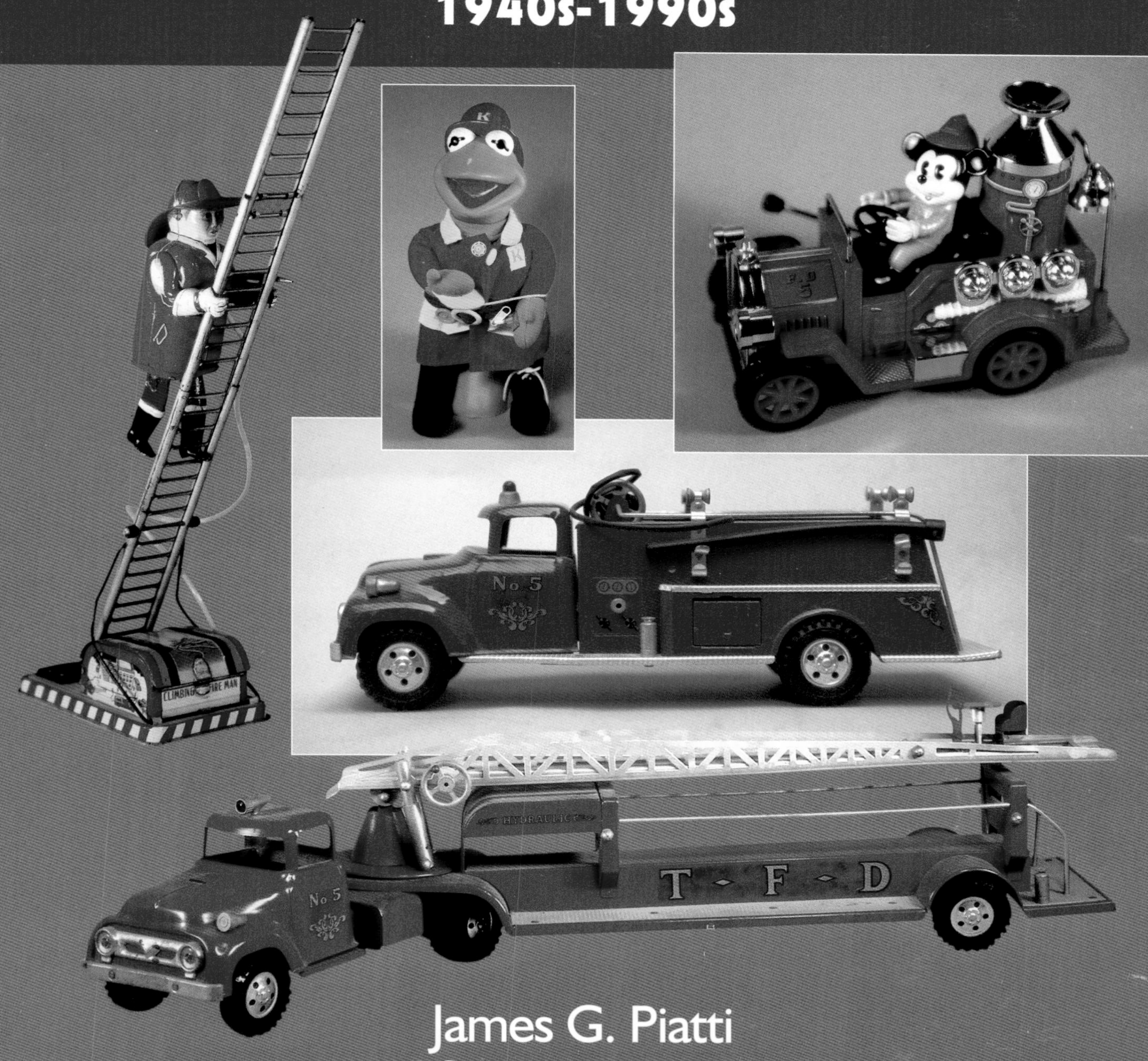


James G. Piatti
and Sandra Frost Piatti

4880 Lower Valley Road, Atglen, PA 19310 USA

# Dedication

To the joys of childhood ... and to heroes everywhere.

Unknown manufacturer, puzzle, firefighters, c. 1970s.


Library of Congress Control Number: 2004118323



Designed by Mark David Bowyer
Type set in Geometric 231 Hv BT/Humanist 521 BT

ISBN: 0-7643-2177-3
Printed in China
1 2 3 4

Published by Schiffer Publishing Ltd.
4880 Lower Valley Road
Atglen, PA 19310
Phone: (610) 593-1777; Fax: (610) 593-2002
E-mail: Info@schifferbooks.com



In Europe, Schiffer books are distributed by
Bushwood Books
6 Marksbury Ave.
Kew Gardens
Surrey TW9 4JF England
Phone: 44 (0) 20 8392-8585; Fax: 44 (0) 20 8392-9876
E-mail: info@bushwoodbooks.co.uk
Free postage in the U.K., Europe; air mail at cost.

# Contents


Acknowledgments ........ 4

Introduction ........ 5

Collector's Tips ........ 6

Chapter One: Firefighting Toys ........ 8

Chapter Two: Smokey the Bear Collectibles ........ 133

Chapter Three: Firefighting Models ........ 136

Chapter Four: TV Show Collectibles ........ 151

Bibliography ........ 160

# Acknowledgments

Our deepest respect and thanks are rendered here to the special ladies who have helped us with this project:

Liz Jocham ... we could not have prepared the manuscript properly without your help.

Donna Baker, our editor ... to you we offer our most heartfelt thanks for your excellent advice, suggestions, and directions.

# Introduction

"Oh please, Santa, put a big red fire engine under the Christmas tree for me this year." Every child whose parents participate in traditional Christmas festivities must entertain similar thoughts, must have similar sugar plum dreams. Indeed, Christmas memories—perhaps more than some other factors—may be the genesis of many professional and private collections of toys. Often, remembered hopes, wishes, and inevitable disappointments give impetus to the adult realizations of childhood dreams.

As a boy, I admired and respected firefighters. The clang of the fire bell, the whine of the siren, and the hustle and urgency of firefighters off to a fire attracted many of my boyhood friends and me. My desire and that of many other children to be close to firefighters and firefighting has kept toy manufacturers busy over many decades making the firefighting toys that filled us with delight and vitalized our play time with color, shape, movement, and sound.

Herein are the results of a lifetime of my collecting in this specific genre of toys. I do hope that you will enjoy them and, in the process, will remember some long lost dreams of your own.

—Jim Piatti

# Collector's Tips

## Where to Buy

The best place to begin your search for firefighting toys is at local antique shows, shops, toy shows, and firefighting memorabilia shows. As your collection grows, attend various collectible shows such as The Spring Thaw Auction and Sale in Allentown, Pennsylvania, the Allentown Toy Show, Chicago Toy Show, the Firehouse Magazine Show and Flea Market, and various muster-flea markets held around the United States. Firematic toys can sometimes be found at car shows as well. Ordinary local flea markets and garage sales may also be considered. Schedules of major antique and toy shows are published in the *Maine Antique Digest* and *The Antiques and The Arts Weekly*.

Auctions are another major source of firematic toys, but please remember: always attend the preview and examine all items carefully at that time. Set a price in your mind and do not go above it. "Auction fever" can be very costly! Another caution: once you buy an item there are no returns—items are sold "as is."

Another source for firefighting toys is online auctions; notable among these is eBay. Thousands of fire-related items are listed every day. In order to participate, you must sign up and receive a password. Auctions last from three days to two weeks. If an item has a reserve price, you are notified and can bid accordingly. Of course, with online auctions one cannot examine the item personally and this may be a problem. You can, however, e-mail questions, and many sellers will give a three day return privilege. It should be noted that auctions end at a specific time and most action occurs during the last two minutes.

## Values

Establishing values for firefighting toys can be difficult. First, the demand far exceeds the supply. A valuable item is rare, but a rare item is not always valuable. Be aware that abundance usually breeds disdain.

For our purposes here, the highest values assigned represent items found to be in mint condition and in their original boxes, while the lowest values represent items in good condition without original boxes.

## Availability

Certain items are readily available, such as new collectible fire toys. However, for older pieces, some items might be once-in-a-lifetime finds and should be purchased. I am reminded of advice I once received from a famous collector. "In regard to unusual items," he said, "you may have to settle for a higher price and lesser condition for that once-in-a-lifetime find. Should a higher grade item become available, you can always upgrade."

The 1990s created two new types of collecting: the limited edition toy and limited edition model. Corgi, Code 3, and several other companies are producing these collectibles. And, as we have said, new collectibles are readily available, but for a limited period of time.

## Demand

As with all collectible items, certain pieces are "hot." This increases demand and value. The problem is that items can "cool down" and their prices will then reflect this change. The jury is still out on value since they are often collected "mint-in-box" and production figures are not always available. I suggest buying them if you like them. Values and collector interest may be undetermined until some future time. Buy what you like and can afford. If the value of your collection grows, all the better.

## Condition

Items shown and discussed in this book range from above average to mint-in-box condition. As a rule, items in mint condition may be worth several times more than an item in good condition while mint-in-box items have the highest value of all. An important consideration with firefighting toys is that they were originally *used*, so finding items in mint condition is difficult and finding them mint-in-box is even more difficult!

## Auctions

Buying at an auction, whether online or live, presents difficulties not found in other methods of purchase. In most buying situations, prices are comparatively fixed, whereas at auctions the purchase price of any one item may vary drastically due to many factors: reserve pricing, geographic location, buyer's premium, and the tendency to succumb to "auction fever"—the fear that someone else will buy this item, causing you or another collector to keep on bidding.

## Panic Buying

Panic buying is a bane to all collectors. Here is a common scenario: you find you haven't bought anything in a couple of weeks and you're at a large antique or toy show. Frustration sets in and you buy on impulse. A week later you look at your purchase and shake your head, wondering how you could have been so stupid. One should make every attempt to avoid "panic buying."

### Cross-Over Value

Any toy with a collectible value in two or more fields of collecting has, in a sense, a heightened value. Furthermore, such items often have the greatest value in the collecting area to which they most specifically pertain. For example, toy collectors will pay higher prices for fire-related toys than will fire collectors.

### Original Parts

Pieces should have original parts and finishing. Refinishing and/or the use of reproduction parts will most often decrease an item's value. Original finish indicating that an item has been used is preferred.

As a final note, this book is a price guide and should be used to assist you in determining values. I am available for questions and comments about this book and continue to be interested in purchasing fire toy collectibles. Please contact Jim Piatti at P.O. Box 244, Oakland, New Jersey 07436. I can also be reached at 973-962-6470.

Chapter One

# Firefighting Toys

The following is a brief history of some twenty-six toy manufacturers—those that were especially popular, and influential throughout the fifty year period featured in this book.

**AMF**: Noted for its magnificent "AMF ROADMASTER" bicycles, the American Machine and Foundry Co. was based in New York City on Madison Avenue. Especially notable to the firefighting toy collector were their well-made, pressed steel fire engine pedal cars and chief's pedal cars. Other sturdy toys produced by AMF were wagons, tractors, trucks, trikes, and bikes.

**AUBURN**: One of the premiere and largest manufacturers of rubber toys, the Auburn Rubber Company of Auburn, Indiana produced rubber toys from approximately 1935-1955. Although the rubber toys were produced for a brief span of only twenty years, their production, which began with a single vehicle (a now highly collectible Cord automobile), was expanded to include trucks, farm tractors, and a myriad of wheeled vehicles. Among these vehicles are fire apparatus. 1952 is said to be the last year of the company's comprehensive rubber toy production. It must be noted that their 1956 toy catalogue lists rubber fire engines; however, all of the other toy vehicles listed are vinyl. Auburn Rubber Company closed its doors in 1969.

**AURORA**: First producing tiny items such as bottle caps and snap together beads, Aurora Plastics Corporation found its golden opportunity when it produced its first model airplane for sixty cents, thus challenging its two dollar and forty-nine cent competitor. Aurora was an innovator. It was the first to print its prices right on the box and the first to use shrink-wrapped packages to prevent loss of pieces. With these improvements, Aurora was able to interest mass-market outlets and retail stores in its toy model kits. Aurora's sales soon skyrocketed. From the 1950s through the 1970s Aurora's business was booming. Its model kits included (besides fire engines), trucks, military vehicles, airplanes, and a complete line of character figures of every sort.

**BUDDY L**: Sometime around the year 1913, an astute businessman named Fred Lundahl started a company that manufactured steel auto and truck parts. When his little son, "Buddy L," reached the age to appreciate large "boy" toys, dad made a few especially for him. Soon Buddy's friends wanted some too. Dad complied. Thus the Buddy L toy company was begun. Made of heavy gauge steel, Buddy L toys could virtually last a lifetime. Also, they were large, somewhere between 21 to 24 inches long for trucks and fire engines. These extremely sturdy toys were made until the 1930s, when a lesser gauge of steel was used by new company owners. The Buddy L company is still in business today.

**CORGI**: Corgi Toys, a division of Mettoy Playcraft Ltd. (previously Playcraft Toys Ltd.), are British made. In 1956, fire engines and some other vehicles first started to be made. It is said, however, that Mettoy (which merged with Playcraft after 1956) had previously issued toys in both metal and plastic. Corgi Toys are especially known for their small (often described as miniature), beautifully detailed and brightly colored, die-cast vehicles, some of which have moving parts. They are still in business today.

**DINKY**: Dinky Toys are of English origin and were manufactured first in the early 1930s under the name "modeled miniatures." Actually, Dinky toys were intended as accessories to model train set-ups. In time, they became popular as toys in their own right. Noted for charming detail, bright colors, and smoothly functioning moving parts, Dinky had its fair share of the die-cast miniature vehicle market. Dinky Toys and Corgi Toys were competitors. Dinky is no longer in business.

**DOEPKE**: The Charles William Doepke Manufacturing Company specialized in toy vehicles that were sturdy beyond all others. Made of superior weight pressed steel, they were large and strong enough to take outdoor play among sand and rocks. Each toy was a highly detailed smaller copy of the original vehicle—down to the realistic tread on the tires! Of major interest to fire toy collectors are the American La France Pumper Fire Truck and the American La France Aerial Ladder Truck. The company went out of business sometime in the mid 1950s.

**FISHER-PRICE**: With an investment of $100,000, the Fisher-Price toy company was begun in the latter part of 1930. It is said that when its founder, Herman G. Fisher, met Irving R. Price, things just fell together. The two men, Fisher and Price, had aesthetic concerns for their toys. They wanted them to be attractively colored and pleasingly shaped. Furthermore, their company creed stated that the toys must have: intrinsic play value, ingenuity, strong construction, good value for the money, and, action. Their earliest toys were pull toys, wind-up, walking, and pop-up toys. By the late 1930s they had begun making Walt Disney authorized toys and from the 1940s to 1960 or so they added musical, pre-school, and educational toys. Plastic parts, added in the 1950s, made production of many items, such as fire engines, cheaper and easier. They are still in business today.

**HESS** (Hess Promotional Toys): In 1958, the Amerada Hess Corp. entered the retail gasoline market and by 1965 there were many stations operating under the brand name of "Hess." In 1964, a toy promotional was begun. A highly detailed and very inexpensive toy tanker truck was offered for sale at Hess gas stations. The truck's fine detail and good price made it an almost instant success. Every year since, a toy has been offered for sale during the holiday season. Fire toy collectors were delighted with the 1970 promotional, as it was a pumper-style fire truck with detachable hoses and ladder. Made in Hong Kong by Marx, it has a revolving red emergency light. In many years thereafter, fire trucks have also been offered—all with excellent detail and life-like features. These trucks are well collected.

**HUBLEY**: The Hubley Manufacturing Company was founded in the late 1800s by John E. Hubley. This Lancaster, Pennsylvania company produced rugged and handsome cast iron children's toys. Toys such as horse-drawn wagons, fire engines, circus trains, and cap pistols were big sellers. Some of these continued to be produced through the Depression. In fact, the "Hubley Fire Department," consisting of pumper, ladder truck, motorcycle, and Fire Chief's car continued to be popular through the end of the Depression years. World War II and its metal restrictions caused toy production to cease temporarily. Post World War II toys were made of die-cast zinc alloy or plastic. The Hubley name ceased to exist sometime in the 1970s. Today, demand is great among collectors for the old cast-iron Hubley toys.

**IDEAL**: Many of us may remember the symbol of the Ideal Novelty & Toy Company—a simple oval with the word IDEAL printed within. As I recall from my childhood in the 1940s and 1950s, that symbol meant "fun" and quality toys. In fact, by the 1950s, Ideal had been in business for almost fifty years, having been founded in 1903 in Brooklyn, New York. Ideal originally specialized in stuffed toys and dolls, however, by the 1940s, plastic, metal, and mechanical toys were being made as well. Firefighting toys were, of course, among them.

**JAPANESE TIN**: Those of us of a "certain age" will remember when Japanese tin toys were synonymous with "cheap," "poorly made," and "dangerous for children," as the sharp, raw edges of these toys could easily draw blood on tender, young fingers. By the 1950s, however, the quality of tin toys had greatly improved. So improved was the quality, in fact, that this era is noted by some toy aficionados to be "the Golden Age" of Japanese tin toys. These prized Japanese tin fire engines and fire chief cars are often found among the inventories of firefighting toy collectors.

**KEYSTONE**: Founded sometime in the early 1920s, Keystone Manufacturing Company's line of toys was perhaps an unusual assortment. Composed of toy motion picture machines, children's movies and cameras, wooden forts and garages, Keystone was, happily, also able to produce some desirable steel fire engines. Each fire truck had fine detail, with additional hoses, ladders, and bells as features. In 1929, the company's finest truck, the 29 inch long water tower, had a brass, pressure-operated water tank (which shot water 25 to 35 feet), brass railing, aluminum running board, brass bell, horn, and balloon tires, and cost a mere eight dollars and forty cents. Needless to say, it is highly prized today. Keystone Manufacturing Company stopped toy production in 1942.

**MARX**: Louis Marx, partnered with his brother, was the founder of the Marx toy empire. Beginning his company in the first decade of the 1900s, Marx originally found his niche by developing the ideas of others and having other firms do the manufacturing. Within the next two decades, Marx and Company began manufacturing their own toys, and a specialty then developed: metal and mechanical toys. It is said that the genius of Louis Marx lay in his ability to capitalize on already popular ideas and to make them cheaper than the competition and in great quantity. The Marx metal fire toys of the 1930s through the 1950s are notable to the collector of fire toys.

**MATCHBOX**: Matchbox toy vehicles were originally made by an English company (Lesney Products), founded in 1947. Very small die cast toys started to become their specialty shortly after the production, in 1953, of their well-remembered mini coronation coach, which commemorated the coronation of Queen Elizabeth II. Fire toy collectors prize the company's tiny, well-made fire vehicles. The company was sold several times and is now owned by Mattel, which continues to produce a line of Matchbox vehicles.

**MATTEL**: One of the most cleverly innovative American toy companies to be presented is Mattel, Incorporated. The Hot Wheels miniature vehicles line, one of their many lines of toys, was quickly successful. Yes, it might be said that Lesney Company's "Matchbox" miniatures paved the way for this triumph, but by adding a racing track, attractive packaging, and varied vehicle selection, Mattel's Hot Wheels caught the imagination of many young folks. Fire engines and fire chief cars are included in the Hot Wheels collection.

**MIDGETOY**: Special circumstances brought about the development of the Midgetoy toy line. The confluence of slack times after World War II and the existence of a tool and gauge company (A&E Tool and Gauge Company) with little work brought about a sturdy line of die-cast vehicle toys. The vehicles often had an Art Deco look and featured a low slung hidden-tire appearance until after the 1970s. Fortunately for fire toy collectors, Midgetoys included fire engines in their toy line. The company went out of business in 1985.

**NYLINT**: A combination of the names Bernard C. Klint and David Nyberg produced the call name for this toy company (Ny-Lint Tool and Manufacturing Company). Formed in 1937 by the pair, the company's early toys were most often wind-up or metal mechanicals. In the beginning of the 1950s, however, large-scale steel trucks and road construction vehicles were featured toys. They also manufactured and still manufacture many bright red fire engines, which, no doubt, have brought much joy to many children.

**PLASTICVILLE**: Bachman Brothers, Incorporated is especially known for its Plasticville line of toys. Total villages and vehicles of plastic, some with moving parts, were manufactured cheaply of plastic for use with train sets. Good pricing made them available to all, and many "Boomer" age people have fond memories of their Plasticville toy sets. Fire toy collectors are especially fond of the fire house and fire engine. They are out of business today.

**REMCO**; Remco Industries, Incorporated, happily for fire toy collectors, made several fire fighting toy vehicles. While the company was primarily noted for battery operated and mechanical toys, some of Remco's other toys were a delight to children: loud speaker trucks and walkie talkies, especially, come to mind.

**REVELL**: Founded in the early 1950s, Revell made good use of new plastics. Scale model kits of a wide range of vehicles were marketed. These model kits, which featured tractors, cars, and fire engines, taught skills and provided hours of constructive fun. Revell went out of business at the end of the 1970s.

**SMITH MILLER**: Founded in 1945, this manufacturer of cast metal trucks and tractors presented two models of its toy vehicles: one was low priced and the other was higher priced and exact in scale and fine in every detail. These larger models are highly collectible and the aerial ladder fire truck is especially well made. The company is still in business today.

**STRUCTO**: Founded in 1908, the Structo Manufacturing Company morphed through several stages of development. Originally, their toys were construction kits similar to those of the A. C. Gilbert Company. Then, early in 1920, pressed steel autos were manufactured. The line was further enlarged to include trucks, riding toys, airplanes, and steam shovels. By 1934, "a pumping fire engine" was being promoted. After World War II, both plastic and steel vehicles were being made.

**TONKA**: Tonka Toys began very humbly in the late 1940s. Featuring pressed steel and authentic and detailed features, Tonka produced a variety of vehicles—among them were fire engines. Early on, the detail on the front ends of their toy trucks was very specific; after 1961, however, truck grills became known as "generic." Tonka is still in business today but most toys are of plastic construction.

**TOOTSIE TOY**: One might say that the Tootsie Toy company began almost by accident. The founding company (named Dowst and Company), was publisher of a trade journal. After their purchase of a linotype machine (in 1893) that was able to produce small metal novelties, the company's production of a small toy limousine and then a tiny toy Ford "fliver" began their toy trade. In 1924, the name Tootsie Toy was patented and though they have been through many changes in ownership since, the company is still alive today. Tiny cast metal fire engines were produced early on. Throughout the years, the size of fire apparatus has ranged from approximately 3 to12 inches. Tootsie toys are highly collectible.

**WYANDOTTE**: The All-Metal Products Company came to be known as "Wyandotte" in perhaps two ways. Some say it was through a clever advertising campaign, others say it was from the company's penchant for printing "Wyandotte" on the sides of some of its vehicles. Founded in 1921, Wyandotte toys flourished through the 1930s. Their lines of cars, trucks, pistols, target games, and doll carriages were often made of sturdy pressed steel and had attractive features such as baked enamel and colorful lithograph design. In the 1950s, Wyandotte was very successful with a heavy-gauge "Rider Fire Truck," made especially appealing by its flashing red light on the roof and a realistic sounding siren. The company came to an end somewhere in the mid-1950s.

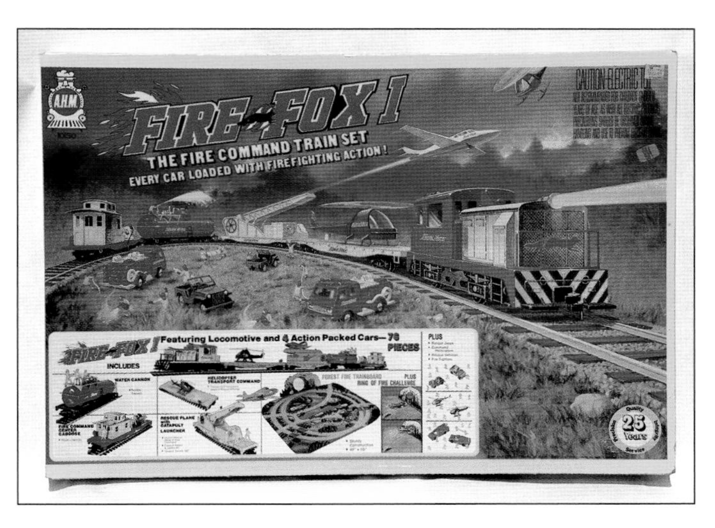


AHM Fire Fox I, Fire Command train set with controls, c. 1970s. $25-$125.

AHM. Fire Fox II, Fire Command train set, plastic with controls, c. 1970s. $25-$125.

AMF firefighter unit No. 508, pedal car, pressed steel, 42", c. 1960s. $100-$250.

AMF hook and ladder, pedal car, pressed steel, 40", c. 1950s. $125-$350.

AMF Wren-Mac, Texaco fire chief fire engine, pressed steel, 25", c. 1960s. $50-$275.

AMF Wren-Mac, Texaco fire chief fire engine box, cardboard, c. 1960s. $50-$100 for box.

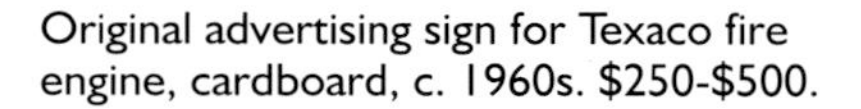

Original advertising sign for Texaco fire engine, cardboard, c. 1960s. $250-$500.

AHI fire chief's car, plastic, c. 1970s. $5-$15.

AHI Walt Disney L'il Zips fire engine, plastic, c. 1970s. $5-$20.

Aladdin, dome top, Disney firefighters lunch box, metal, 9", c. 1960s. $25-$150.

American Thermos, dome top, fire station lunch box, metal, 9", c. 1950s. $50-$275.

American Thermos, Thermos bottle for fire station lunch box, steel, 8" tall, 1950s. $30-$50.

Arco, Disney Donald Duck fire truck, die cast, c. 1970s. $5-$15.

Arco, Disney Mickey Mouse fire truck, die cast, c. 1970s. $5-$15.

Arco, Mickey's Fire Truck, plastic, c. 1980s. $5-$10.

Arco, Disney Fire Truck Play Set, plastic, c. 1970s. $5-$15.

Auburn Rubber, Ahrens Fox fire engine, rubber, rare, 5.5", c. 1940s. $25-$175.

Auburn Rubber, rescue squad fire engine, rubber, 5.25", c. 1960s. $5-$15.

Aurora, AFX fire chief slot car, plastic, rare, c. 1970s. $15-$75.

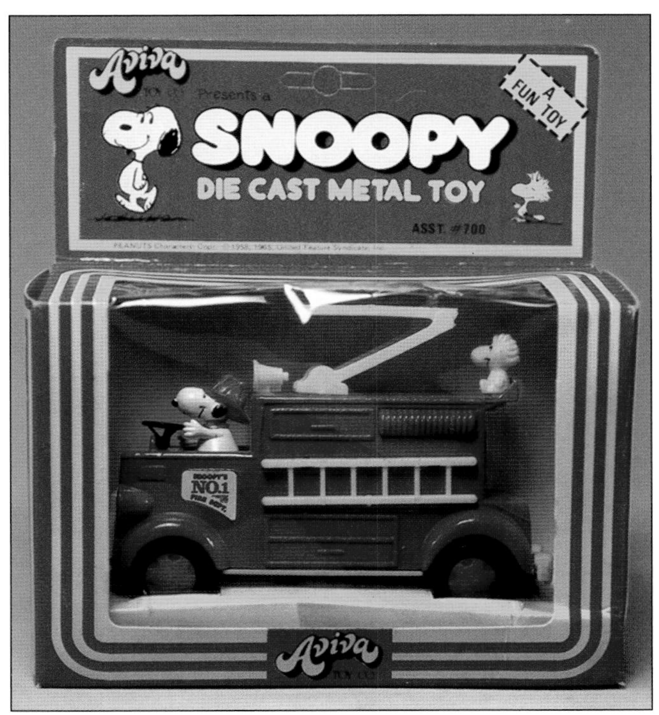


Aviva, Snoopy fire engine, No. 200, die cast, c. 1960s. $10-$35.

Aviva, Snoopy's Pin Pals fireman, plastic, c. 1970s. $2-$5.

Aviva, Snoopy fire engine, die cast, c. 1970s. $5-$15.

Aviva, Snoopy fire engine, No. 72222, die cast, c. 1970s. $5-$10.

Banner Plastics Corp., hook & ladder, plastic, c. 1950s. $5-$15.
Unknown maker, fire house, metal, c. 1950s. $75-$175.

Bachmann, Plasticville fire house, plastic with two plastic fire engines, c. 1960s. $5-$25.

Left to right: Barkley fireman in parade position, 3.5"; H. O. fireman with hose, 1.25"; Barkley fireman with hose, 3", all antimonial lead, c. 1950s-60s. L: $15-$30; C: $5-$15; R: $15-$30.

Banthrico, steam fire engine bank, cast metal, 3.75", c. 1960s. $15-$45.

Buddy L, GMC tractor drawn hook and ladder, pressed steel, c. 1950s. $125-$300.

Buddy L, Ford rescue van, No. 3, pressed steel, c. 1960s. $150-$350.

Buddy L, aerial ladder fire engine, pressed steel, c. 1960s. $75-$200.

Buddy L, Ford pumper, No. 3, pressed steel, 15", c. 1960s. $75-$200.

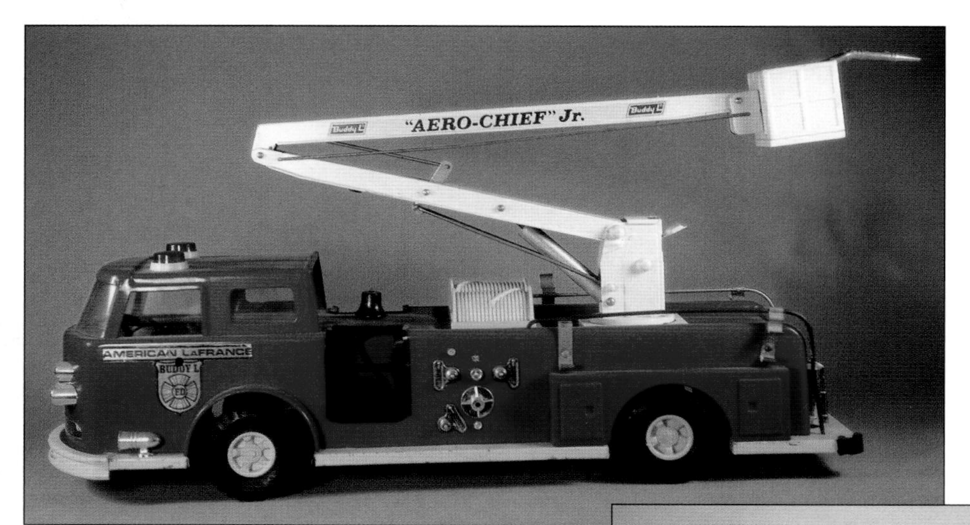


Buddy L, Aero-Chief Jr. by American LaFrance, pressed steel, 29", c. 1960s. $125-$300.

Buddy L, hydraulic snorkel pumper, pressed steel, c. 1960s. $50-$125.

Buddy L, pumper, pressed steel, c. 1960s. $15-$35.

Buddy L, searchlight fire truck with battery-operated light, pressed steel, c. 1960s. $25-$60.

Buddy L, hook-n-ladder, pressed steel, c. 1960s. $25-$75.

Buddy L, snorkel, pressed steel, c. 1960s. $10-$35.

Buddy L, open cab, American LaFrance snorkel, pressed steel, c. 1960s. $25-$150.

Buddy L, Jr. Brutes fire pumper, pressed steel, c. 1970s. $5-$15.

Buddy L, fire emergency pumper, pressed steel, c. 1970s. $5-$25.

Buddy L, Lil Buddy's fire pumper, pressed steel, c. 1970s. $5-$15.

Buddy L (left), open cab pumper; (right) open cab hook and ladder, pressed steel, c 1970s. L: $5-$15; R: $8-$20.

Buddy L, fire chief's car, pressed steel, c. 1970s. $5-$15.

Buddy L, ladder truck, 6", open cab, pressed steel and plastic, c. 1970s. $5-$20.

Buddy L, ladder truck, closed cab, pressed steel and plastic, c. 1970s. $5-$20.

Buddy L, fire emergency rescue set, pressed steel and plastic, c. 1970s. $15-$35.

Buddy L, Lil Brutes, fire pumper, pressed steel, c. 1970s. $5-$15.

Buddy L, Mack trucks, hook and ladder, pressed steel, c. 1970s. $10-$35.

Buddy L, hook and ladder, pressed steel, c. 1970s. $8-$15.

Buddy L, fire pumper with fireman and dog, pressed steel and plastic, c. 1980s. $10-$30.

Buddy L, hook and ladder, No. 4901, pressed steel, c. 1970s. $15-$35.

Buddy L, electronic chief's car, pressed steel, battery operated, c. 1980s. $5-$10.

Buddy L, electronic fire rescue truck, battery operated, c. 1980s. $5-$10.

Buddy L, Steel Stormers, airport foam unit, pressed steel, c. 1980s. $5-$10.

Buddy L, Lil Buddy's, fireman, plastic, c. 1960s. $15-$35.

Buddy L, fire department set, pressed steel and plastic, c. 1980s. $10-$35.

Burger King giveaway, fire engine, plastic, 2.5", c. 1970s. $2-$5.

Childcraft, fire chief's car, wood, 12", c. 1950s. $25-$95.

Child Guidance, Sesame Street firehouse, plastic. c. 1970s. $15-$35.

Coleco, Cabbage Patch Kids, firefighter outfit, c. 1980s. $5-$10.

Coleco, Flintstone Kids Bedrock fire pumper, plastic, c. 1980s. $10-$20.

Commonwealth Plastics Corp., fire chief siren, plastic, c. 1950s. $25-$50.

Corgi, Major, American LaFrance, aerial rescue truck, die cast, c. 1970s. $25-$125.

Corgi, Chevy Impala fire chief's car, die cast, c. 1960s. $25-$100.

Corgi, Simon Snorkel, die cast, c. 1970s. $25-$200.

Corgi, HCB-Angus, Firestreak, Dodge truck, die cast, c. 1980s. $25-$150.

Corgi, fire engine, die cast, c. 1980s. $15-$75.

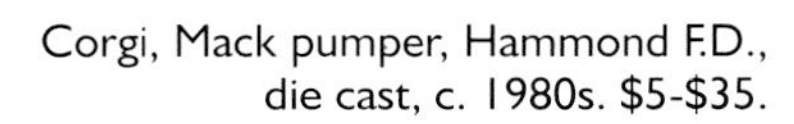
Corgi, Mack pumper, Hammond F.D., die cast, c. 1980s. $5-$35.

Corgi, Cub fire chief's car, die cast, c. 1970s. $5-$35.

Corgi, Corgitronics fire chief, No. 1008, die cast, c. 1980s. $10-$35.

Corgi, Junior, fire chief's car, No. 70, die cast, c. 1970s. $5-$15.

Corgi, Junior, fire boat, No. 53. $5-$10.

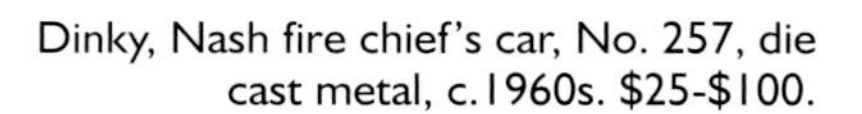

Dinky, Nash fire chief's car, No. 257, die cast metal, c.1960s. $25-$100.

Dinky, fire chief's car, No. 195, die cast metal,
c. 1970s. $25-$90.

Dinky, E.R.F. fire tender, No. 266, die cast metal, c. 1970s. $25-$100.

Dinky, fire engine with extension ladder, No. 955,
die cast metal, c. 1970s. $25-$75.

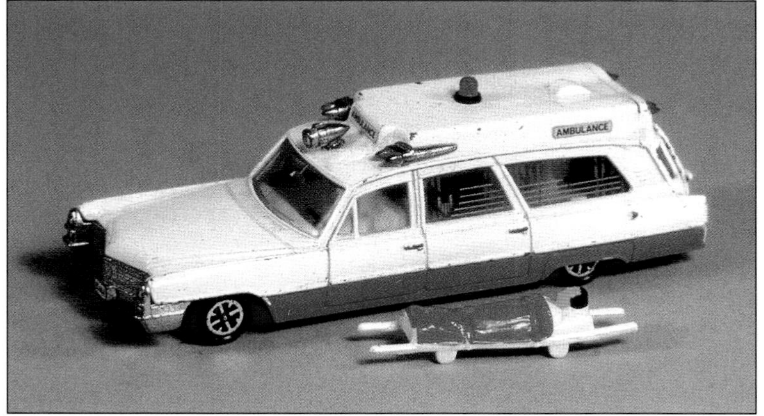


Dinky, Superior Cadillac ambulance, No. 288, die
cast metal, 5.75", c. 1970s. $20-$50.

Wm. Doepke Mfg. Co., "Model Toys," American LaFrance ladder truck, No. 2014, pressed steel, 33", c. 1950s. $100-$350.

Wm. Doepke Mfg. Co., ad from 1955 Christmas toy catalog. $50.

Wm. Doepke Mfg. Co., American LaFrance pumper, No. 2010, 19", pressed steel, c. 1950s. $100-$450.

Wm. Doepke Mfg. Co., searchlight truck, No. 2010, exceedingly rare, pressed steel, c. 1950s. $500-$2,000.

Duplo, Lego fire chief's car, No. 2403, plastic, c. 1970s. $10-$25.

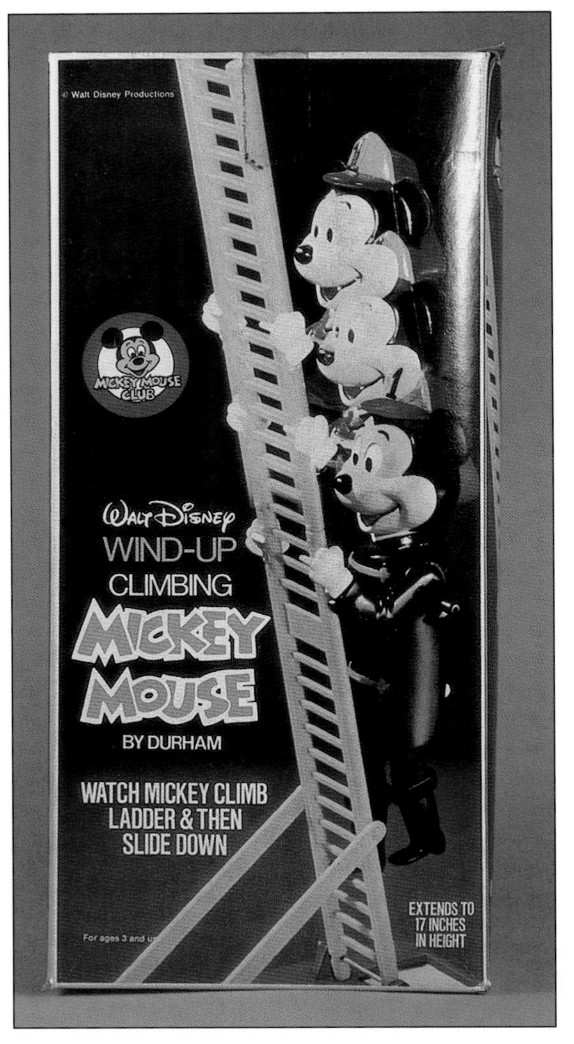


Durham, wind-up climbing Mickey House, plastic, c. 1970s. $25-$125.

Durham, climbing Mickey Mouse, No. 1530, plastic, c. 1970s. $10-$35.

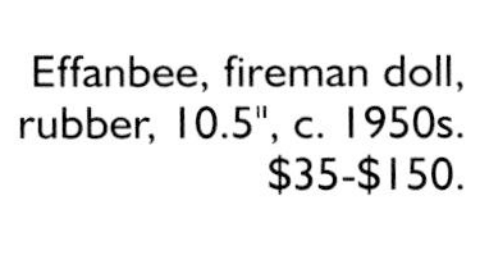

Effanbee, fireman doll, rubber, 10.5", c. 1950s. $35-$150.

Eldon, Big Poly Toys, Seagrave pumper, plastic, c. 1950s. $25-$175.

Electric Game Company, Jim Prentice, electric firefighting game, c. 1950s. $50-$100.

Electric Game Company, Jim Prentice, electric firefighting game board.

Ertl, Daffy Duck firefighter, die cast, c. 1990s. $5-$10.

Ertl, fire truck, No. 9 play set, die cast and plastic, c. 1980s. $25-$75.

Ertl, replica fire and rescue, Pierce pumper, die cast, c. 1990s. $5-$10.

Ertl, Smurf fire truck, die cast, 1980s. $5-$25.

Ertl, replica fire and rescue, Pierce aerial fire truck, die cast, c. 1990s. $5-$10.

F.A.O. Schwartz firehouse, wood, 16" x 18.5", c. 1940s-1950s; Kay-Dee Plastics, fire truck, plastic, 9.5", c. 1940s. Firehouse: $50-$300; Fire truck: $10-$35.

Fisher-Price, Winky Blinky fire truck, wood, c. 1950s. $75-$150.

Fisher-Price, fire truck, No. 1920, wood and plastic, c. 1960s. $5-$20.

Fisher-Price, fire truck, No. 124, plastic, c. 1970s. $5-$25.

Fisher-Price, Little People fire truck, No. 0124, plastic, c. 1985. $5-$15.

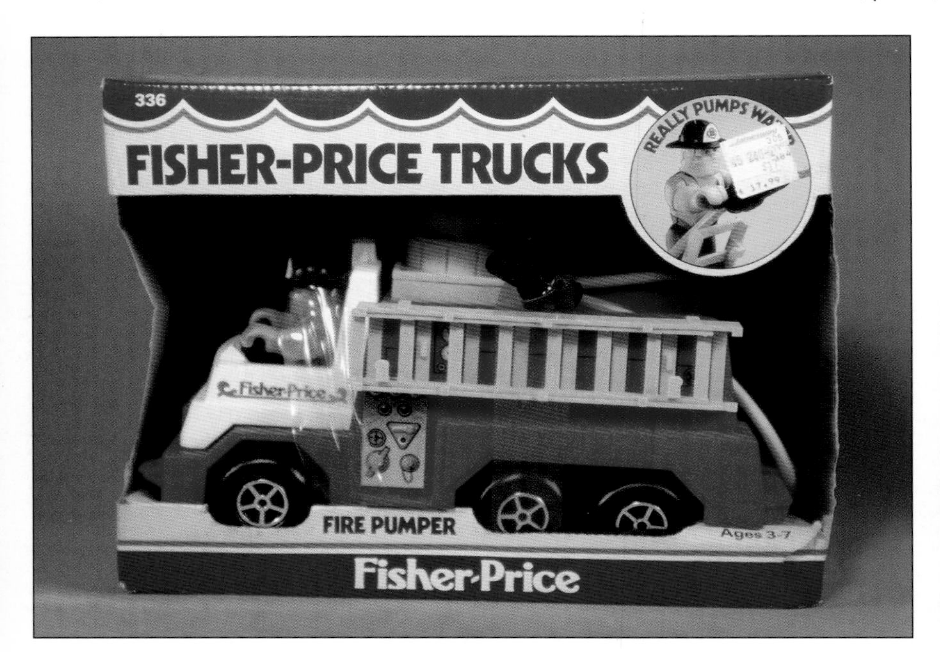


Fisher-Price, fire pumper, No. 336, plastic, c. 1983. $5-$25.

Fisher-Price, Look and Play books, *A Visit to the Firehouse*, plastic firefighter, c. 1980s. $5-$15.

Fisher-Price, hook and ladder, No. 319, plastic, c. 1983. $5-$25.

Fisher-Price, Little People, fire truck set, No. 10346, plastic, c. 1985. $5-$20.

Fisher-Price, Adventure People, paramedic, No. 383, plastic, 1980s. $5-$25.

Fisher-Price, firefighter set, No. 2004, plastic, c. 1985. $10-$50.

Fisher-Price, Crazy Clown Fire Brigade, No. 657, plastic, c. 1983. $10-$35.

Fisher-Price, Classics, fire engine, No. 2250, plastic, 1990s. $5-$25.

Funrise, Micro Action fire engines, plastic, c. 1989. $3 individual, $25 set.

GIJ Toys, World of Bendables, firefighter, plastic, c. 1970s. $5-$25.

Gabriel, fire engine, die cast, c. 1960s. $8-$25.

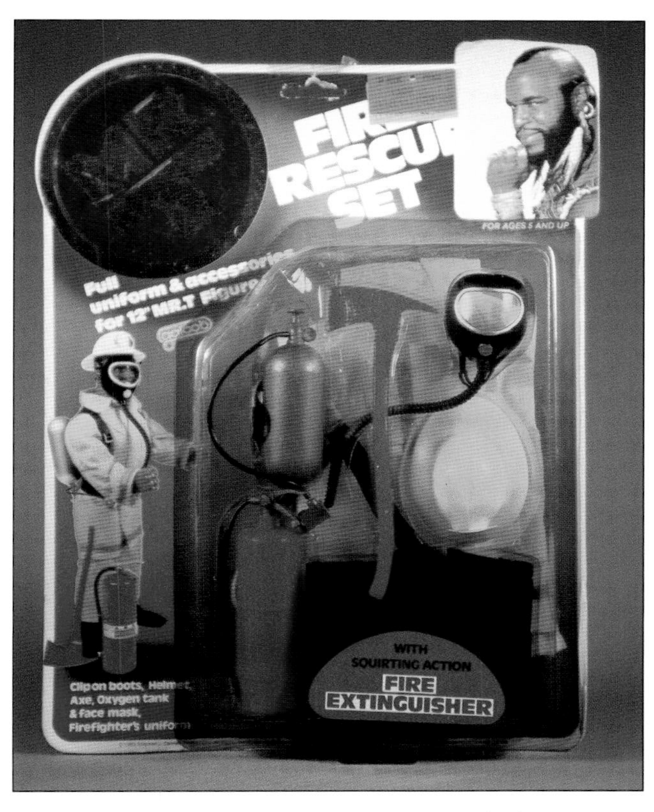


Galoob, Mr. "T" fire rescue set, plastic, c. 1983. $5-$35.

Galoob, Micro Machines, fire station 13 play set, plastic, 1990s. $5-$35.

Galoob, Micro Machines, #1 Rescue Squad, plastic, 1990s. $5-$25.

Galoob, Micro Machines, #5 Firefighters, plastic, 1990s. $5-$25.

Galoob, Micro Machines, City Scenes, fire station, plastic, 1990s. $5-$25.

Galoob, Micro Machines, #6, Blaze and Roar, plastic, c. 1990s. $5-$25.

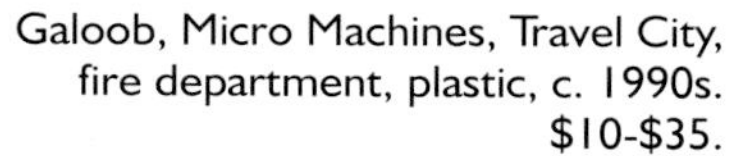

Galoob, Micro Machines, Travel City, fire department, plastic, c. 1990s. $10-$35.

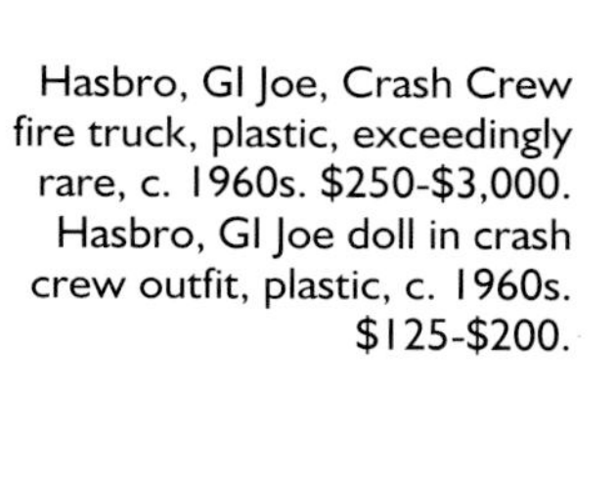

Hasbro, GI Joe, Crash Crew fire truck, plastic, exceedingly rare, c. 1960s. $250-$3,000. Hasbro, GI Joe doll in crash crew outfit, plastic, c. 1960s. $125-$200.

Hasbro, GI Joe Action Pack, firefighter, plastic, c. 1980s. $25-$175.

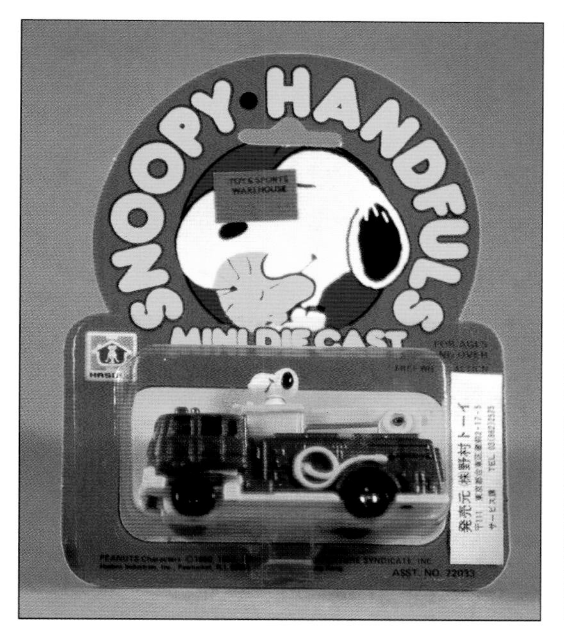


Hasbro, Snoopy Handfulls, fire engine, die cast, c. 1980s. $3-$15.

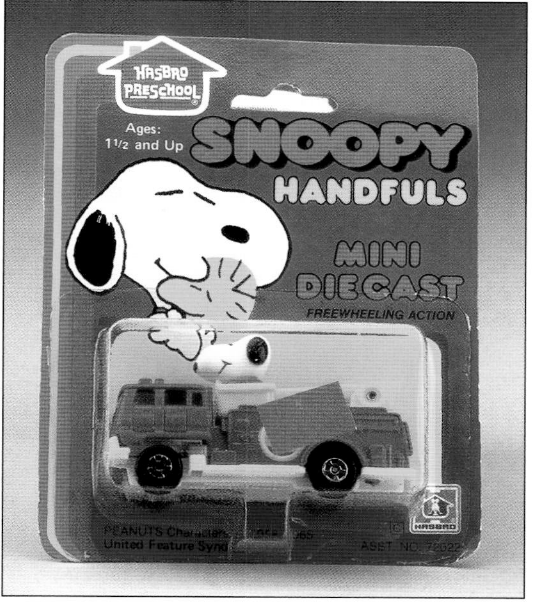


Hasbro, Snoopy Handfuls, fire engine, die cast, c. 1980s. $3-$15.

Hasbro, Transformers, Autobot Search and Rescue Inferno, die cast, c. 1980s. $5-$50.

Hassenfield Bro., Fearless Fireman game, c. 1957. $20-$100.

Hassenfield Bro., Fearless Fireman game, interior view.

Hess, fire truck, plastic, c. 1970. $35-$500.

Hess, advertisement for 1986 fire truck, plastic. $75.

Hess, box for fire truck, c. 1986. $15-$25.

Hess, fire truck, plastic, 14", c. 1986. $15-$75.

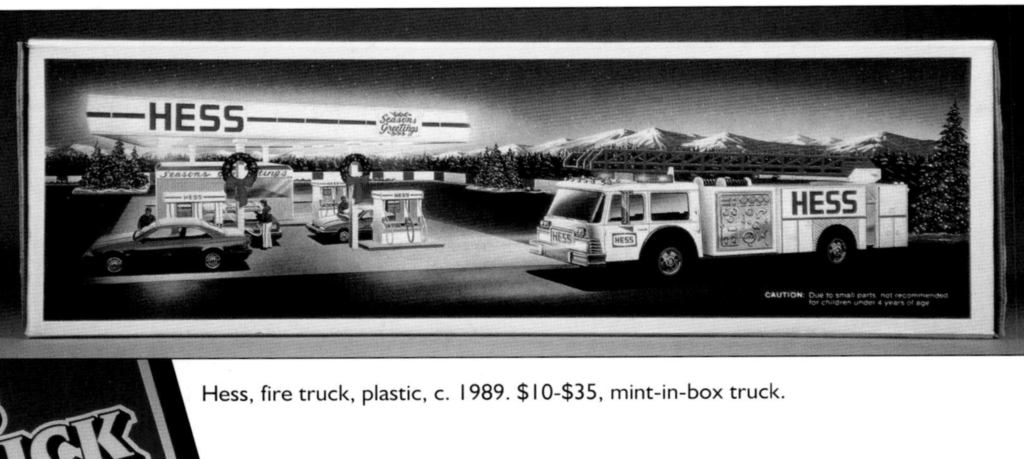


Hess, fire truck, plastic, c. 1989. $10-$35, mint-in-box truck.

Hubley, Ahrens-Fox fire engine, die cast, 7.5", c. 1940s. $35-$125.

Hess, advertisement for 1989 fire truck. $50.

Hubley, Ahrens-Fox closed cab fire engine, No. 953, cast metal, 7.5", c. 1948. $15-$125.

Hubley, rare box for Ahrens-Fox closed cab fire engine, No. 953. $75.

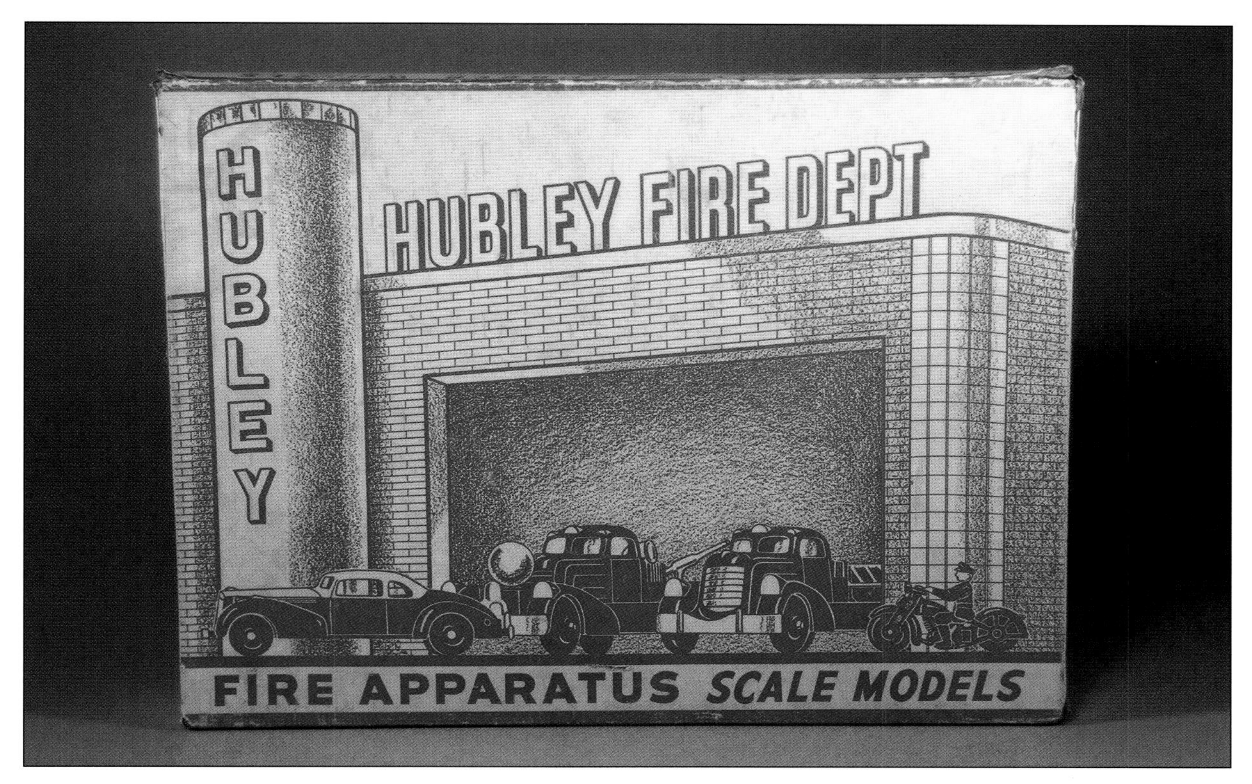


Hubley, fire apparatus scale models set, cast metal, c. 1948. $250-$500.

Hubley, hook and ladder, No. 472, cast metal, 10.25", c. 1948. $20-$75.

Hubley, fire engine, cast metal, 11.25", c. 1940s. $15-$45.

Hubley, fire engine, No. 464, cast metal, 7.5", c. 1940s. $25-$150.

Hubley, fire engine, No. 520, cast metal, 16.5", c. 1956/57. $75-$200.

Hubley, fire engine, No. 333, plastic, 8.5", c. 1956. $15-$35.

Hubley (front), fire engine, plastic, No. 306, 5.75"; (rear) hook and ladder, plastic, No. 307, 5.75", c. 1956/57. $15-$30 each.

Hubley, American LaFrance fire engine, No. 513, cast metal, 19", c. 1956, rare. $50-$200.

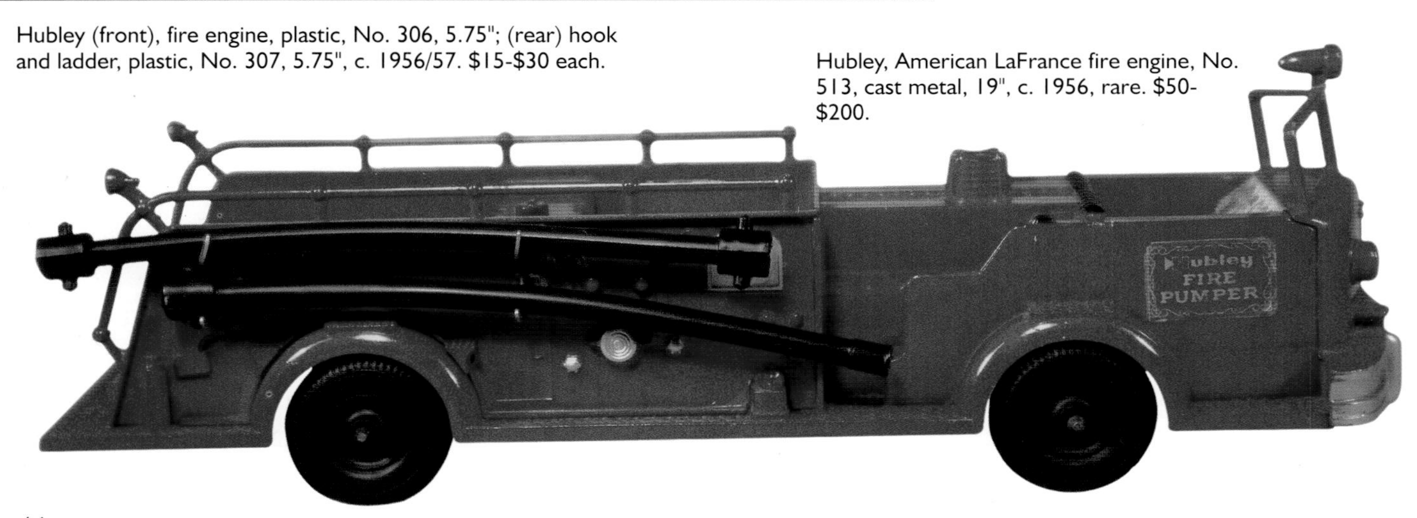

Hubley, hook and ladder, No. 471, cast metal, c. 1957/64. $35-$150.

Hubley, hook and ladder, No. 472, cast metal, 11.25", c. 1957-1964. $25-$150.

Hubley, Forest Ranger Helicopter, No. 483, cast metal, 9.5", c. 1959-62. $25-$100.

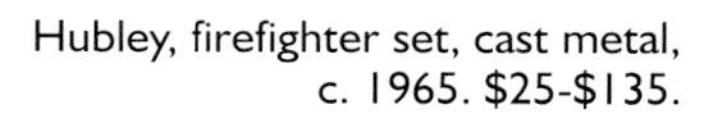

Hubley, firefighter set, cast metal, c. 1965. $25-$135.

Hubley, rescue truck, No. 1493, cast metal and plastic, 8.25", c. 1965. $25-$100.

Hubley, American LaFrance pumper, cast metal, 6", c. 1962-65. $15-$25.

Hubley, Tiny Toy kit, American LaFance pumper, plastic, 4.5", c. 1960s. $5-$15.

Hubley, Tiny Toy kit, American LaFrance ladder truck, plastic, 6.25", c. 1960s. $5-$15.

Ideal, water pumping fire boat, plastic, c. 1960s. $15-$125.

Ideal, Motorific Trucks, fire engine, plastic, c. 1960s. $25-$100.

Ideal, Motorific Boats, fire boat, plastic, c. 1960s. $25-$100.

Ideal, Tiny Mighty Mo, fire engine, plastic, c.1960s. $5-$75.

Ideal, Lighted Slam Shifters, chief's car, No. 48163, die cast, c. 1970s. $5-$25.

ILLCO, Bert, Gyro Fire Truck, plastic. c. 1980s. $5-$30.

ILLCO, Bubble Blowing Fire Truck, plastic, c. 1980s. $5-$45.

ILLCO, Mickey Mouse Gyro fire engine, plastic, c. 1970s. $15-$75.

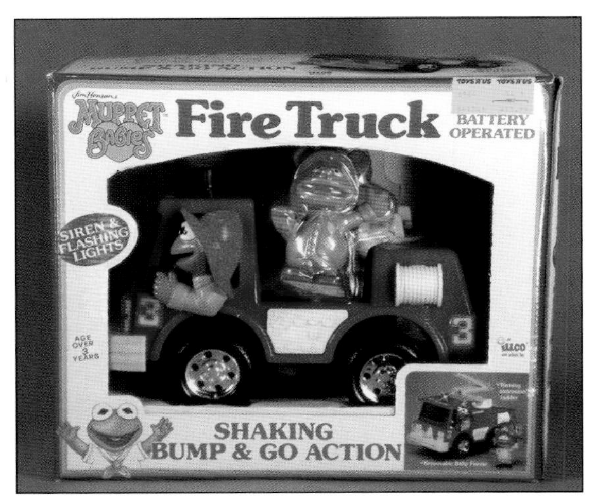


ILLCO, Muppet Babies Fire Truck, Shaking Bump & Go Action, plastic, c. 1970s. $5-$25.

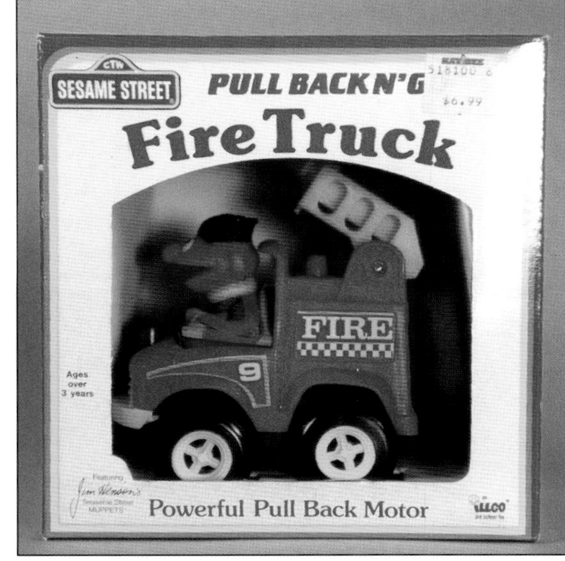


ILLCO, Sesame Street, Pull Back N'Go fire truck, plastic, c. 1970s. $5-$35.

International Training Tech, Snoopy fire engine, battery operated, plastic, c. 1970s. $5-$35.

International Training Tech, Snoopy fire engine, cast metal, c. 1970s. $5-$35.

IPS Limited, climbing fireman, plastic, c. 1980s. $5-$20.

Its-A-Beaut, fire chief's car, cast metal, 4", c. 1940s. $10-$30.

Japanese, manufacturer unknown, Jeep fire engine, tin, c. 1950s. $15-$45.

Japanese, manufacturer unknown, fire engine, tin, c. 1950s. $25-$85.

Japanese, manufacturer unknown, "Old Smoky" with smoke, tin, c. 1950s. $75-$250.

Japanese, manufacturer unknown, friction powered fire engine with bell ringing, tin, c. 1960s. $25-$125.

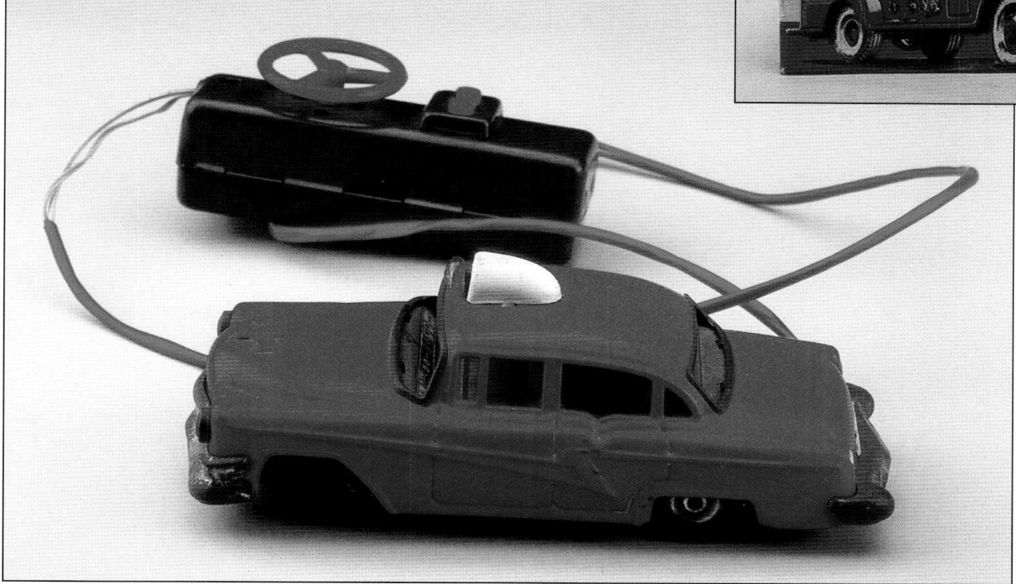

Japanese, manufacturer unknown, fire chief's car, battery operated, tin, c. 1950s. $25-$100.

Japanese, manufacturer unknown, hook and ladder, tin, 14", c. 1950s. $15-$65.

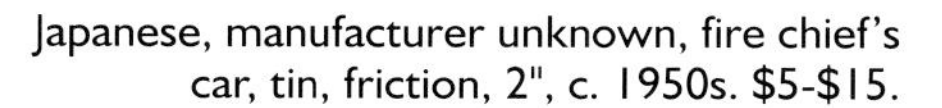

Japanese, manufacturer unknown, fire chief's car, tin, friction, 2", c. 1950s. $5-$15.

Japanese, manufacturer unknown, fire chief's car, tin, friction, 8", 1950s. $8-$35.

Japanese, manufacturer unknown, fire chief's car, tin, friction, 8", c. 1960s. $8-$35.

Japanese, manufacturer unknown, fire boat, No. 3500, tin, battery operated, 15", c. 1960s. $75-$200.

Japanese, manufacturer unknown, old fashioned fire engine (part of a set), tin, friction, 7", c. 1960s. $15-$35.

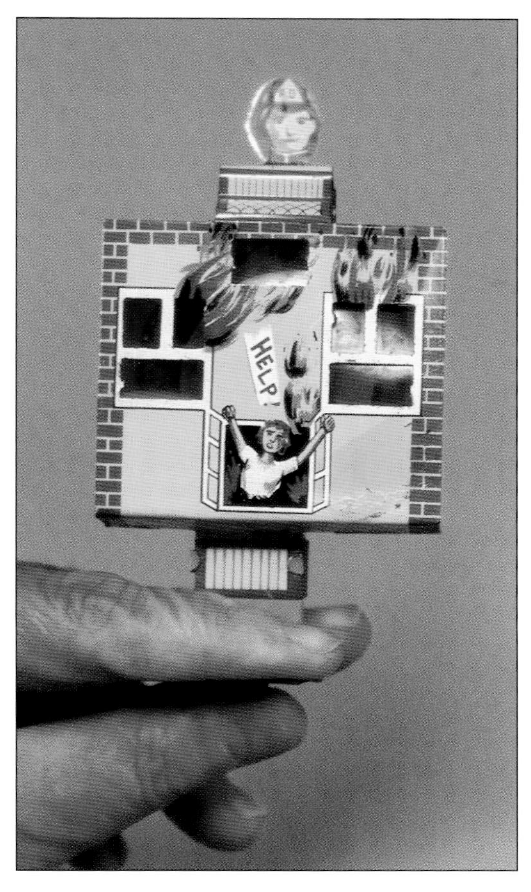

Japanese, manufacturer unknown, fireman siren, tin, c. 1940s. $10-$35.

Japanese, manufacturer unknown, burning building sparkler, tin, c. 1950s. $25-$75.

Japanese, manufacturer unknown, climbing fireman, tin (battery operated), 15.5", c. 1950s. $75-$175.

Japanese, manufacturer unknown, fire chief's car, tin, battery operated, 10", c. 1960s. $35-$100.

Jaymar, puzzle, Currier & Ives, New York fire scene, c. 1970s. $5-$25.

Jamar, puzzle, Campbell Kids fire engine, c. 1980s. $15-$50.

Jamar, puzzle, Disneyland fire engine, c. 1970s. $10-$25.

Johnson and Johnson, Peg Pals at the Firehouse, cardboard and plastic, c. 1980s. $5-$15.

Jet, fire boat, plastic, c. 1980s. $10-$35.

Kellogg, Frosted Flakes fire engine, giveaway, plastic, c. 1960s. $5-$10 each.

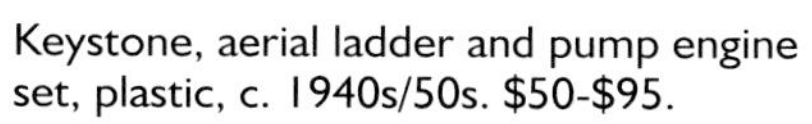

Keystone, aerial ladder and pump engine set, plastic, c. 1940s/50s. $50-$95.

Keystone, two bay fire station, pressed board, 16" long x 8" high, c. 1950s. $50-$125.

Keystone, fire boat, wood, 12", c. 1940s/50s. $25-$100.

Keystone, three bay firehouse, rare flat roof, pressed board, c. 1950s. $75-$150.

Knickerbocker, fire boat, plastic, c. 1950s. $10-$75.

Kidco, Tough Wheels, fire chief's car, cast metal, c. 1980s. $5-$25.

Left: Kohner, Spinning Barney, plastic fireman, 4.75", c. 1960s; right: Unknown, Fred Flintstone, plastic, 1.75", c. 1960s. L: $10-$25; R: $5-$10.

Lapin, fire chief's car, plastic, 5.75", c. 1950s. $8-$35.

Lido Toy Corp., fire engine, plastic, 4.75", c. 1940s/50s. $5-$15.

Lionel, fire car, No. 52, c. 1950s. $100-$250.

Lionel, fire fighting instruction car, No. 6530, c. 1960s. $50-$200.

Lionel, ladder car, No. 3512, c. 1950s. $75-$200.

Lucky Plastic Co., ladder truck, plastic, c. 1960s. $25-$95.

Madame Alexander, female firefighter doll, c. 1990s. $50-$100.

Majorette, Sonic Flashers, fire department SUV, cast metal, c. 1980s. $3-$15.

Majorette, Sonic Flashers, fire boat, plastic, c. 1980s. $5-$20.

Majorette, Sonic Flashers, fire engine, cast metal, c. 1980s. $3-$15.

Majorette, Micro Sonic Flashers, fire chief's car, cast metal, c. 1980s. $2-$10.

Majorette, Super Movers, hook and ladder, cast metal, c. 1980s. $5-$35.

Majorette, Super Movers, fire rescue set, die cast, c. 1980s. $15-$50.

Majorette, Super Movers, rescue squad, die cast, c. 1980s. $5-$35.

Majorette, fire engine, No. 207, die cast, c. 1980s. $3-$10.

Majorette, fire engine, No. 3000, die cast, c. 1980s. $5-$35.

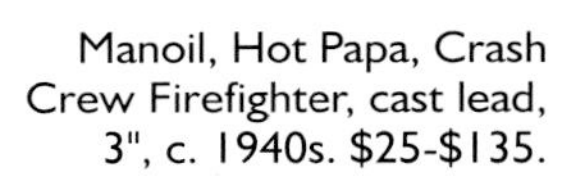
Manoil, Hot Papa, Crash Crew Firefighter, cast lead, 3", c. 1940s. $25-$135.

Manoil, fire engine, slush toy, 2.75", c. 1940s. $15-$40.

Manoil, fire engine, 4", c. 1950s. $8-$35.

Marx, fire chief's car, plastic, c. 1940s. $25-$135.

Marx, ladder truck, friction powered, tin, c. 1950s. $60-$200.

Marx, fire chief's car, key wind, battery, tin, 10.5", c. 1950s. $50-$200.

Marx, fire house play set, tin, rare, c. 1950s. $500-$1500.

Marx, climbing fireman, key wind, tin, 22.5", c. 1950s. $100-$225.

Marx, climbing fireman, key wind, plastic, 22.5", c. 1950s. $50-$150.

Marx, rescue squad, pressed steel, 12.5", c. 1950s. $35-$100.

Matchbox, Dennis fire engine, No. 9, die cast metal, 2.25", c. 1955. $10-$50.

Matchbox (Lesney), fire station, plastic, c. 1960s. $25-$175.

Matchbox (Lesney), Merryweather Marquis fire engine, die cast metal, c. 1950. $10-$50.

Matchbox (Lesney), airport crash tender, die cast metal, c. 1964. $5-$35.

Matchbox (Lesney), fire pumper, No. 29, die cast metal, c. 1965. $8-$35.

Matchbox (Lesney), Ford fire chief's car No. 59, with box, c. 1966. $10-$30.

Matchbox (Lesney), Merryweather fire engine, die cast metal, c. 1970. $8-$35.

Matchbox, Mickey Mouse, firefighter, die cast, c. 1979. $10-$60.

Matchbox (Lesney), top: Blaze Buster, No. 22, die cast metal, c. 1975; bottom: Snorkel fire engine, No. 13, die cast metal, c. 1977. $5-$30 each.

Matchbox, Mobile Action Command, rescue team, c. 1970s. $10-$35.

Matchbox Pre-school, Twinkletown firehouse, plastic, c. 1970s. $5-$20.

Matchbox, Emergency Station, plastic, c. 1970s. $25-$50.

Matchbox (Lesney), fire chief's car, No. 64, die cast metal, c. 1976. $3-$20.

Matchbox (Lesney), Blaze Buster, No. 22, die cast metal, c. 1976. $3-$20.

Matchbox (Lesney), Snorkel fire engine, No. 63, die cast metal, c. 1970s. $3-$30.

Matchbox (Lesney), ladder truck, die cast metal, c. 1980s. $3-$30.

Matchbox (Lesney), airport foam tender, No. 54, die cast metal, c. 1980s. $3-$30.

Matchbox (Lesney), airport foam tender, No. MB 54, die cast metal (red color), c. 1980s. $3-$30.

Matchbox (Lesney), emergency services storage box, plastic, c. 1980s. $5-$15.

Matchbox (Lesney), Super Kings, Snorkel fire engine, No. K-39, die cast, c. 1970s. $10-$95.

Matchbox (Lesney), King Size Merryweather fire engine, No. K-15, die cast, c. 1964. $5-$65.

Matchbox (Lesney), Super Kings, Dodge Monaco fire chief's car, No. K-67, die cast, c. 1970s. $10-$75.

Matchbox (Lesney), Super Kings, Dodge Monaco fire chief's car, Hackensack, N.J., No. K-67, die cast, c. 1970s. $10-$75.

Matchbox (Lesney), Super Kings, airport crash tender, No. K-75, die cast, c. 1970s. $10-$75.

Matchbox, Speed Kings, Hot Fire Engine, No. K-53, die cast, c, 1980s. $10-$50.

Matchbox, Speed Kings, Mercedes ambulance, No. K-63, die cast, c, 1980s. $10-$35.

Matchbox, gift set, Metro Fire Department, die cast, c. 1980s. $15.

Matchbox, Super Kings, Peterbilt fire spotter plane transport, No. K-134, die cast, c. 1980s. $10-$85.

Matchbox, Convoy, hook and ladder, die cast, c. 1980s. $5-$12.

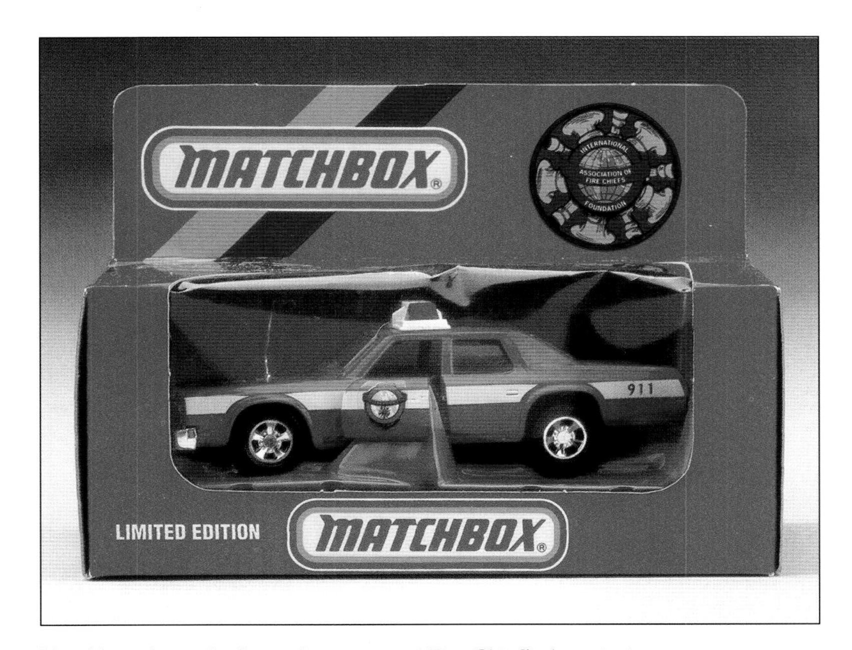


Matchbox, limited edition, International Fire Chief's Association fire chief's car, die cast, c. 1980s. $25-$100.

Matchbox, Super Kings, fire chief's car, No. K 78, die cast, c. 1980s. $10-$35.

Matchbox, Emergency Action Pack, No. 50110, die cast, c. 1980s. $15.

Matchbox, World's Smallest, Mini Playset #1, fire house, die cast, c. 1980s. $12.

Matchbox, Twinkletown, Freddy Ready Fire Truck, c. 1980s. $5-$15.

Mattel, Barbie Caring Careers, firefighter outfit, c. 1990s. $5-$25.

Mattel, Talkin' Tracks Fire Station, plastic, c. 1970s. $10-$50.

Mattel, Hot Wheels, Red Lines, hook and ladder, die cast, c. 1970s. $50-$125.

Mattel, Hot Wheels, Crack-Ups, fire chief's car, No. 565, c. 1970s. $5-$35.

Mattel, Hot Wheels, Flying Colors, Chief's Special, No. 22, die cast, c. 1970s. $5-$15.

Mattel, Hot Wheels, Flying Colors, Fire-eater, No. 11, die cast, c. 1970s. $5-$15.

Mattel, Hot Wheels, Flying Colors, Emergency Squad, No. 23, die cast, c. 1970s. $5-15.

Mattel, Hot Wheels, Flying Colors, fire chief's car, die cast, c. 1970s. $5-$15.

Mattel, Hot Wheels, Rescue Team, Emergency Squad, die cast, c. 1970s. $5-$50.

Mattel, Hot Wheels, Fire Chaser, No. 2639, die cast, c. 1979. $5-$50.

Mattel, Hot Wheels, fire engine, die cast, c. 1970s. $5-$50.

Mattel, Hot Wheels (left), Fire Eater, c. 1970; (right) Old #5, die cast, c. 1981. L: $5-$35; R: $10-$50.

Mattel, Hot Wheels, Work Horses, rescue squad, No. 5145, die cast. $5-$35.

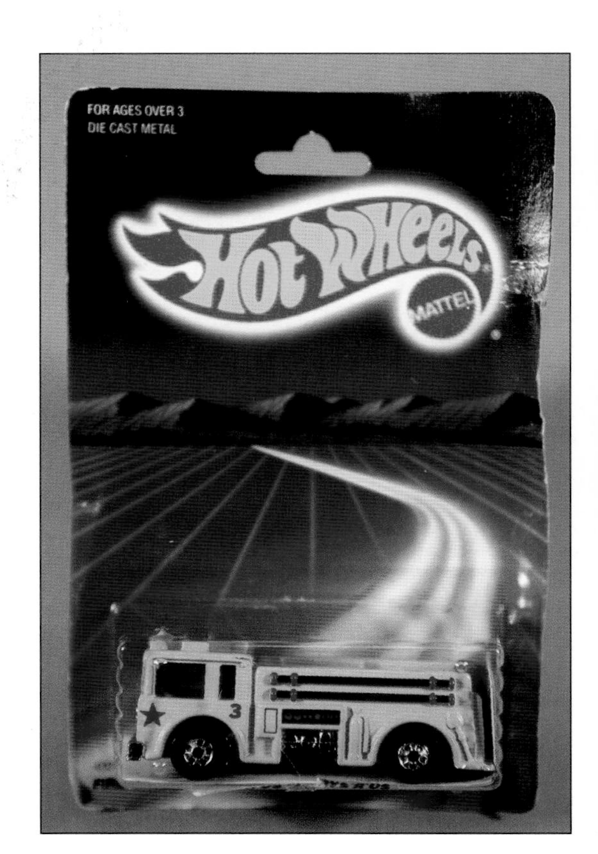


Mattel, Hot Wheels, fire engine, die cast. $5-$20.

Mattel, Hot Wheels, Work Horses, Fire Eater, No. 9640, die cast, c. 1980s. $3-$25.

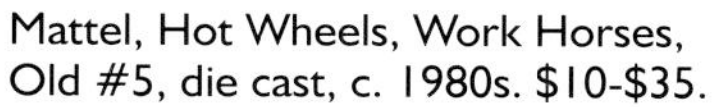

Mattel, Hot Wheels, Work Horses, Old #5, die cast, c. 1980s. $10-$35.

Mattel, Hot Wheels, Work Horses, Flame Stopper, No. 77, die cast. c. 1988. $3-$20.

Mattel, Hot Wheels, Work Horses, Airport Rescue, No. 1699, die cast, c. 1980s. $3-$35.

Mattel, Hot Wheels, firefighter play set, No. 4225, die cast, c. 1980s. $50.

Mattel, Hot Wheels, Firefighter Sto & Go play set, c. 1980s. $75.

Mattel, Men of Medal, fire chief, plastic, c. 1980s. $5-$20.

Mattel, Preschool, Hub-Bubs Mr. Dog, dog firefighter, plastic, c. 1980s. $5-$25 MIP as shown.

Mattel, Disney, fire truck, plastic, c. 1970s. $10-$50.

Mattel, Disney Mickeytown Fire Station play set, plastic, c. 1980s. $50.

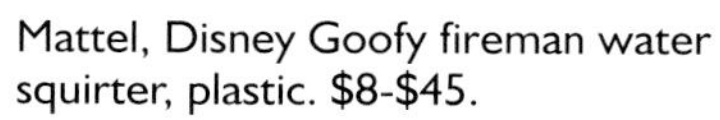
Mattel, Disney Goofy fireman water squirter, plastic. $8-$45.

Mattel, Preschool, First Wheels, fire chief's car, die cast, c. 1980s. $3-$12.

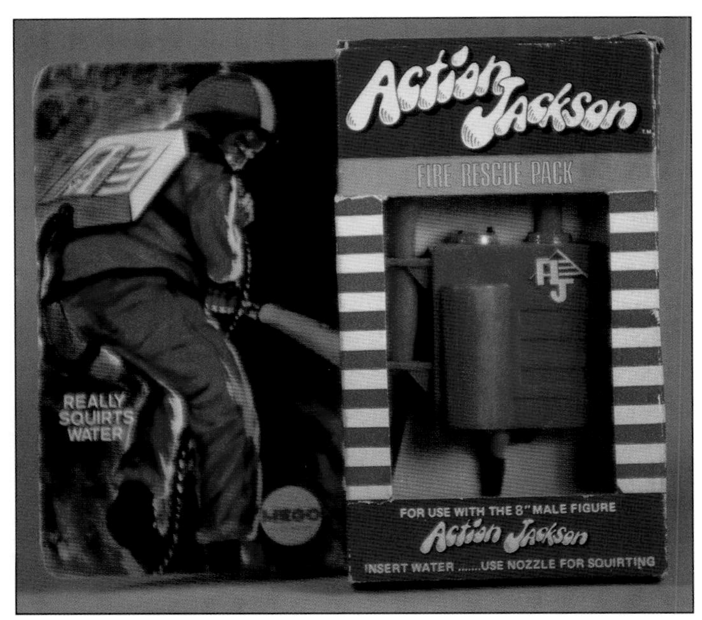


Mego, Action Jackson, fire rescue pack, c. 1974. $5-$30.

McCrory, Free Wheelers, fire department station wagon, die cast, c. 1970s. $5-$18.

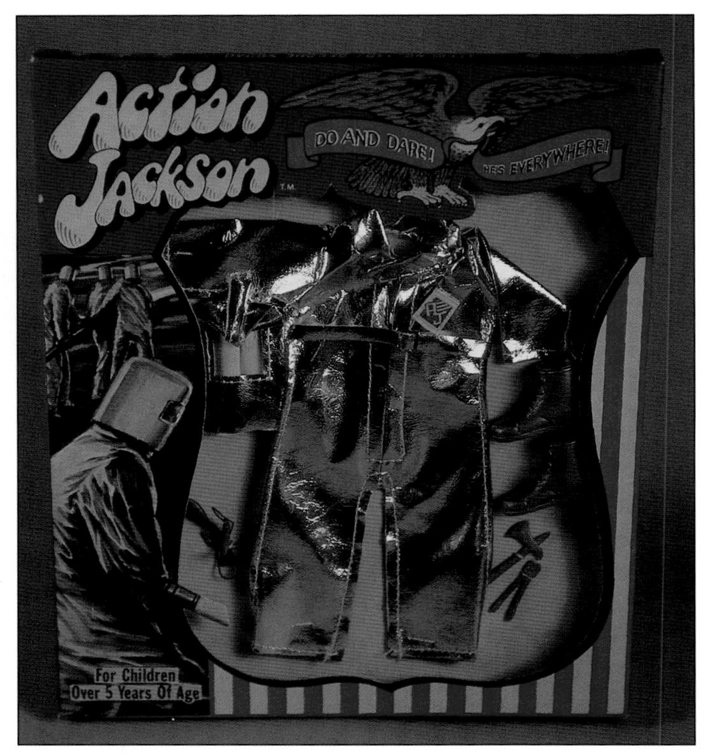


Mego, Action Jackson, crash crew rescue outfit, c. 1974. $8-$35.

Metal Masters, fire engine, cast metal, 10", 10", c. 1940s. $10-$35.

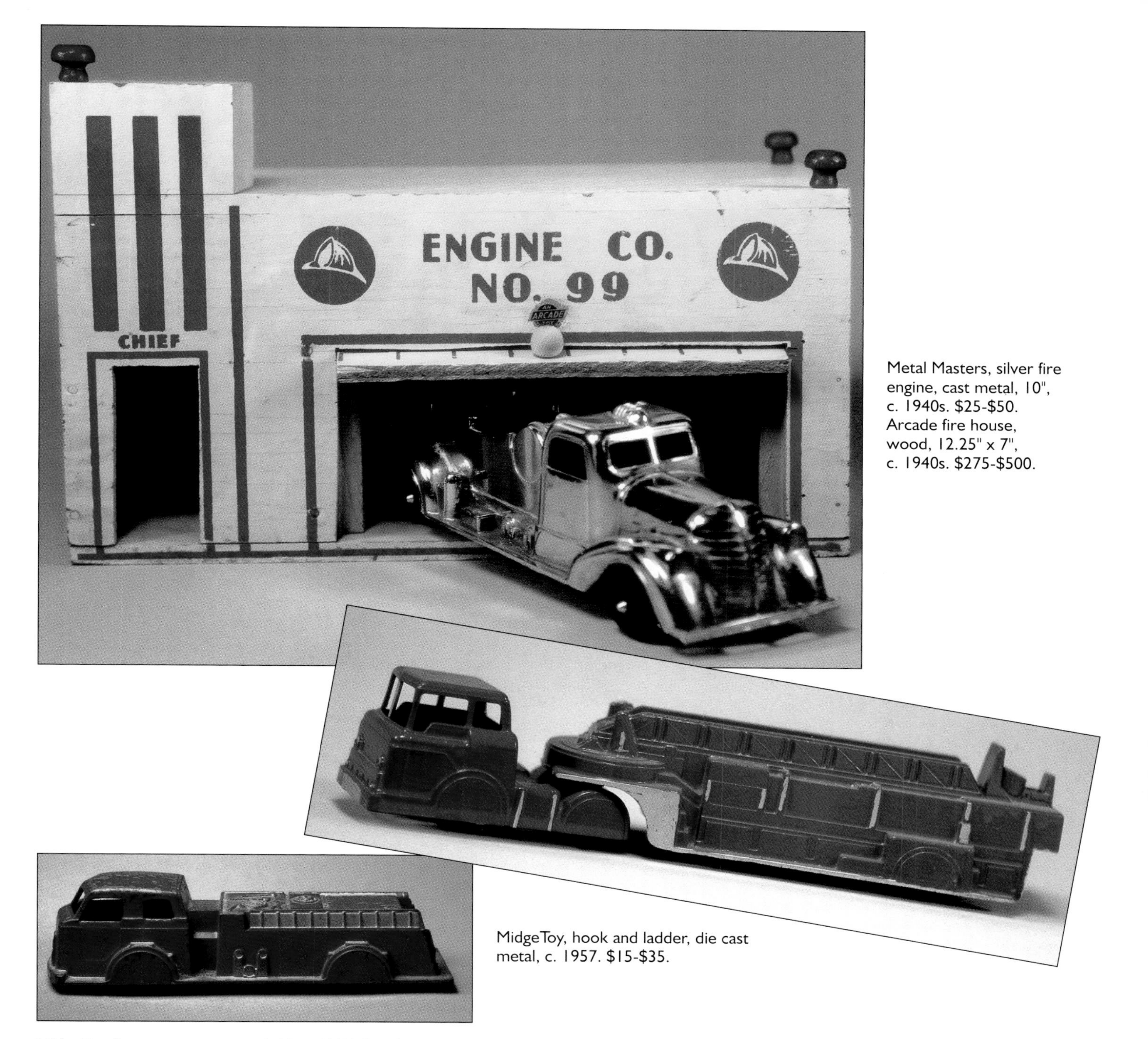


Metal Masters, silver fire engine, cast metal, 10", c. 1940s. $25-$50. Arcade fire house, wood, 12.25" x 7", c. 1940s. $275-$500.

MidgeToy, hook and ladder, die cast metal, c. 1957. $15-$35.

MidgeToy, fire engine, cast metal, 6", c. 1957. $15-$40.

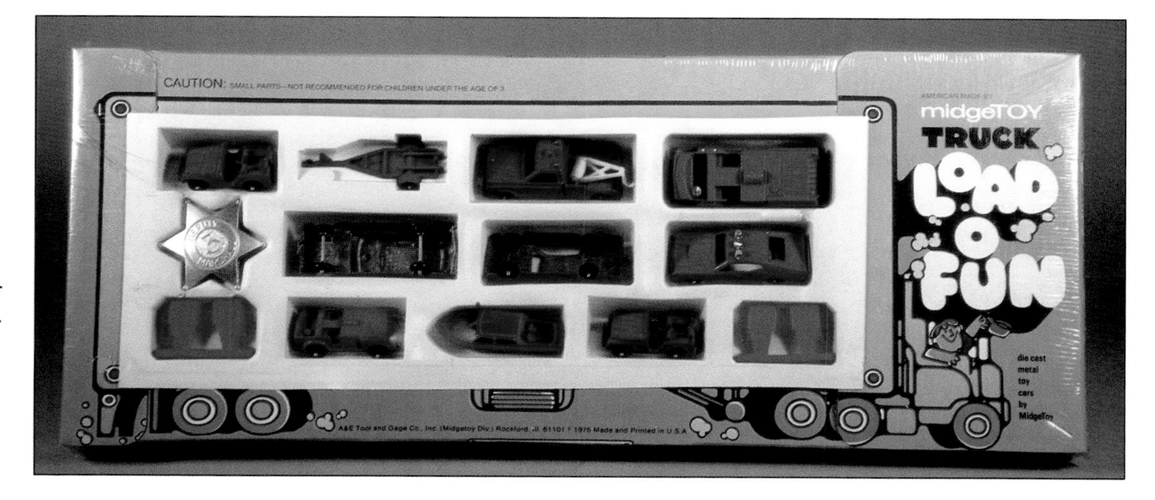


MidgeToy, Load of Fun, firefighter set, die cast metal, c. 1975. $75.

Milton Bradley, Fire Alarm game, c. 1950s. $25-$60.

Milton Bradley, fire engine puzzle, c. 1960s. $5-$15.

Milton Bradley, Dalmatian puzzle, c. 1960s. $5-$15.

Model Power, Mack snorkel, die cast, c. 1980s. $5-$35.

Model Power, American LaFrance, snorkel, die cast, c. 1980s. $8-$35.

Model Power, firefighter's train set, c. 1980s. $95.

Model Power, Road Kings, fire chief's car, die cast, c. 1970s. $3-$20.

Model Power, firefighters HO scale antique chief's car, die cast, c. 1980s. $8-$25.

Model Power, firefighters HO scale firefighting equipment, die cast metal, c. 1980s. $5-$20 each.

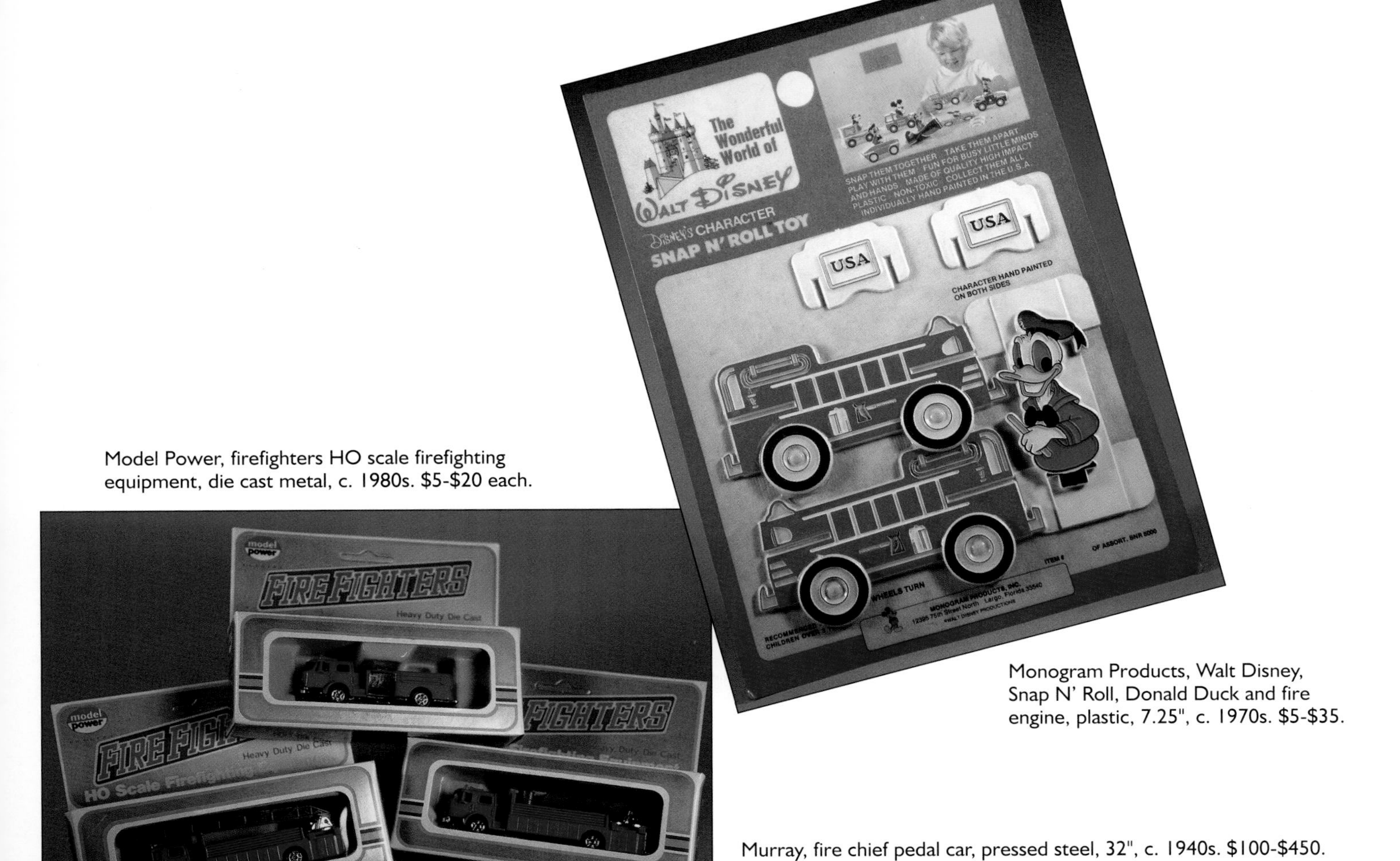


Monogram Products, Walt Disney, Snap N' Roll, Donald Duck and fire engine, plastic, 7.25", c. 1970s. $5-$35.

Murray, fire chief pedal car, pressed steel, 32", c. 1940s. $100-$450.

Murray, fire chief pedal car, pressed steel, 32", c. 1950s. $100-$350.

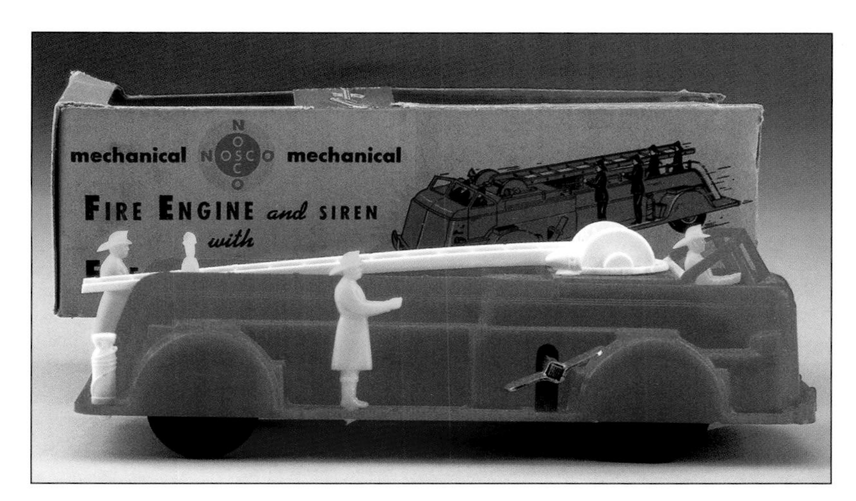


Nasco, hook and ladder, plastic, 8", c. 1940s. $15-$50.

Nasts Industries, Motor Friend, fireman and fire engine, plastic, c. 1960s. $25.

New Bright, Bump-N-Go, fire engine, battery operated, plastic, c. 1970s. $10-$35.

New Bright, fire boat, battery operated, plastic, c. 1970s. $10-$65.

Nylint, pumper, pressed steel, 10", c. 1975. $20-$150.

New Bright, fire engine, The Snorkel Truck, remote controlled, battery operated, plastic, c. 1980s. $20-$50.

Nylint, Ford Bronco, fire chief's car, pressed steel, c. 1960s. $25-$125.

Nylint, emergency truck, pressed steel, c. 1960s. $35-$150.

Nylint, fire chief's car, pressed steel, c. 1960s. $20-$100.

Nylint, snorkel, pressed steel, c. 1960s. $35-$150.

Nylint, rescue pumper, pressed steel, c. 1960s. $20-$100.

Nylint, firehouse set, No. 942, pressed steel, c. 1970s. $30-$200.

Nylint, Aerial Hook-N-Ladder, pressed steel, c. 1960s. $25-$200.

Nylint, Cadet, aerial ladder, pressed steel, c. 1970s. $25-$125.

Nylint, Cadet, snorkel truck, pressed steel, c. 1970s. $20-$150.

Nylint, classic pumper, pressed steel, c. 1980s. $20-$100.

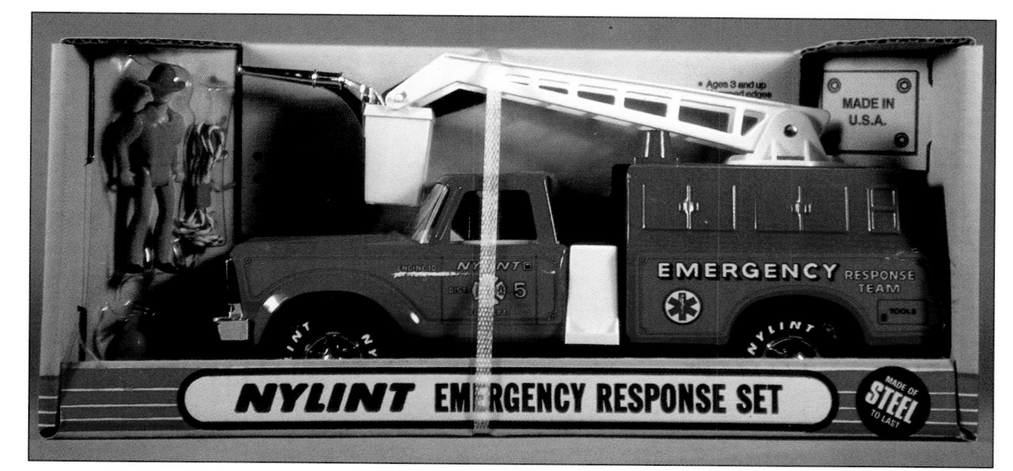


Nylint, Cadet, Emergency Response Set, pressed steel, c. 1970s. $20-$100.

Nylint, rescue pumper, pressed steel, c. 1980s. $20-$100.

Nylint, water cannon, pressed steel, c. 1980s. $25-$125.

Nylint Classics, pumper, pressed steel, c. 1993. $15-$45.

Nylint, Men of Steel, firefighter, plastic, c. 1990s. $3-$18.

Nylint, Metal Muscle, rapid response vehicle, pressed steel, c. 1990s. $10-$35.

Nylint, Metal Muscle, aerial hook-n-ladder, pressed steel, c. 1990s. $15-$75.

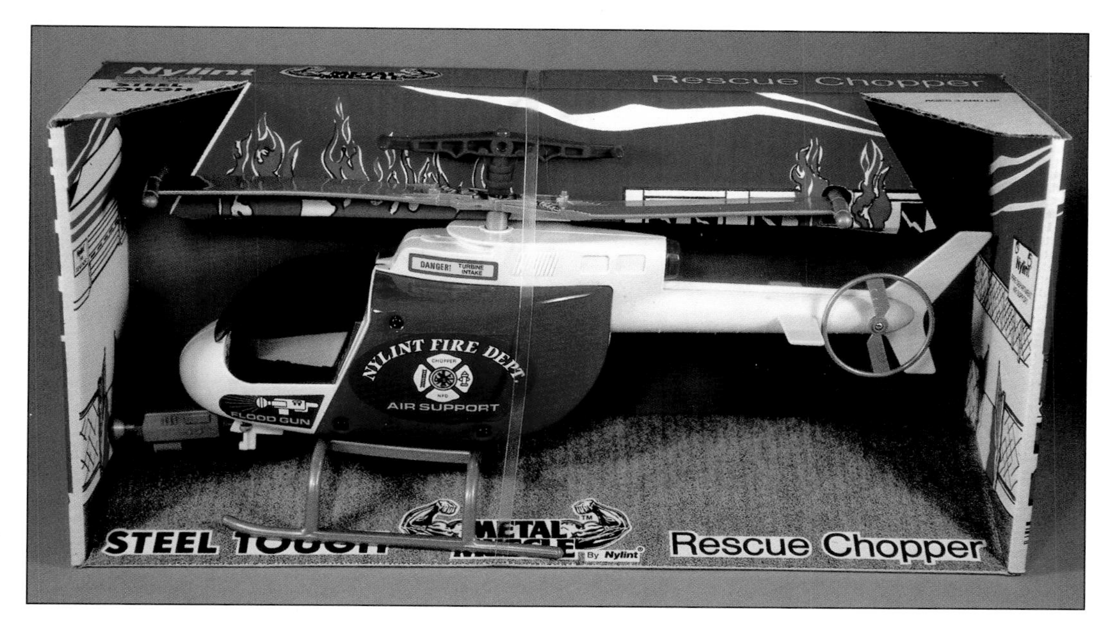


Nylint, Metal Muscle, rescue chopper, pressed steel, c. 1990s. $15-$85.

Nylint, Metal Muscle, Fire Fighters, pumper, pressed steel, c. 1990s. $5-$20.

Nylint, Metal Muscle, Fire Fighters, ladder truck, c. 1990s. $5-$20.

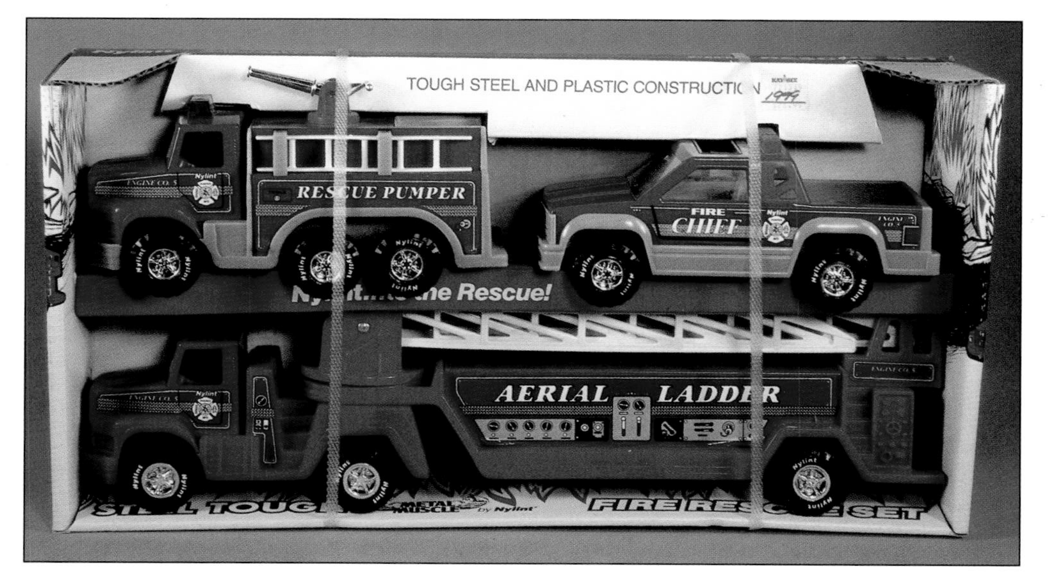


Nylint, Fire Rescue Set, pressed steel and plastic, c. 1990s. $75.

Nylint, Metal Muscle, aerial hook-n-ladder, pressed steel and plastic, c. 1990s. $6-$25.

Nylint, Metal Muscle, rescue pumper, pressed steel, c. 1990s. $5-$25.

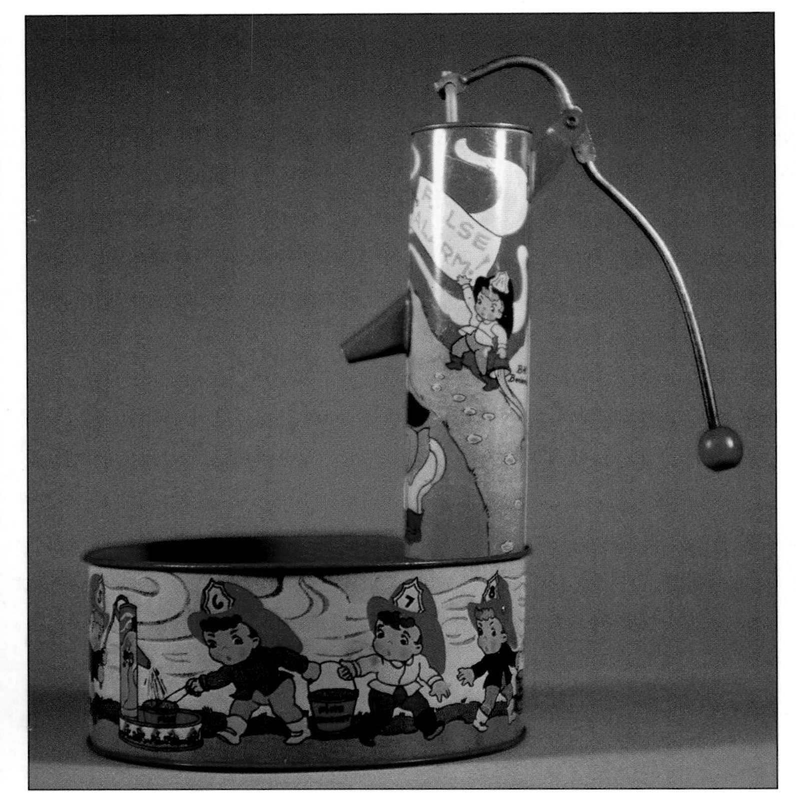

Ohio Art, child's beach toy, firefighter pump, tin, 9.5", c. 1940s. $25-$125.

PlayArt, American LaFrance, pumper, die cast, c. 1980s. $10-$35.

PlayArt, American LaFrance, aerial, die cast, c. 1980s. $10-$35.

PlayArt, Fast Wheel, fire engine, die cast, c. 1970s. $5-$15.

PlayArt, Fast Wheel, fire truck, die cast, c. 1970s. $5-$15.

Playskool, Me and My Buddy, Fireman Adventure Set, c. 1970s. $5-$35.

Playskool, Sesame Street, Ernie's Fire Engine, die cast, c. 1970s. $5-$25.

Pez, candy dispenser (left): with cardboard display packaging, plastic; (right): later issue. L: $100; R: $10-$50.

Processed Plastic, American LaFrance Pumper, 6", plastic, 5.75", c. 1950s. $5-$25.

Polistil, fire rescue ambulance, die cast metal, c. 1980. $20-$80.

Ralstoy, Step Vans, cast metal, 4.5", c. 1970s. $10-$35 each.

Ralstoy, Step Vans, Newark Fire Department, cast metal, 4.5", c. 1970s. $10-$35.

Remco, Heavy Metal, hook and ladder, pressed steel, c. 1980s. $25-$125.

Remco, Heavy Metal, pumper, pressed steel, c. 1980s. $25-$125.

Remco, Heavy Metal, fire department set, pressed steel, c. 1980s. $35.

Remco, Heavy Metal, hook and ladder, pressed steel, c. 1980s. $8-$25.

Remco, Heavy Metal, snorkel, pressed steel, c. 1980s. $5-$25.

Remco, Heavy Metal, hook and ladder, c. 1980s. $5-$35.

Remco, Zybots, robot fire engine, c. 1980s. $5-$35.

Renwal, chief's car, plastic, 4.5", c. 1948-56. $5-$15.

Renwal, fire engine, cast metal, c. 1980s. $25-$100.

Renwal, fire alarm, plastic, 4.5", c. 1950s. $25-$95.

Renwal, fire engine, No. 59, 7", c. 1952-54. $25-$135.

Renwal, fire boat, plastic, c. 1950s. $35-$250.

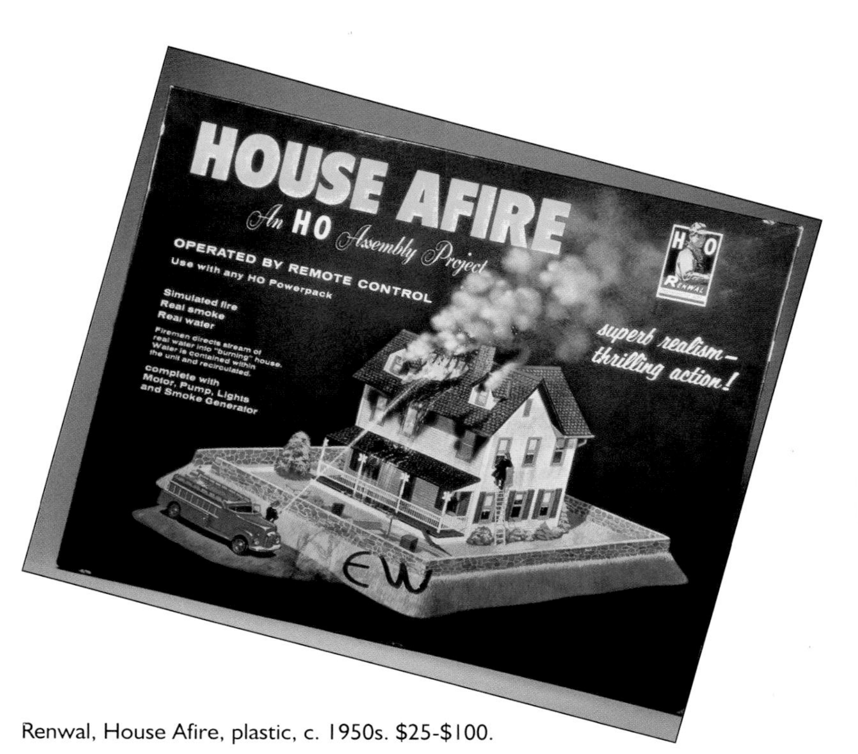


Renwal, House Afire, plastic, c. 1950s. $25-$100.

Roadchamps, American LaFrance snorkel, cast metal, c. 1990. $5-$20.

Road Tough, Emergency Team play set, die cast metal, c. 1990s. $25.

Road Tough, Emergency Team, chief's car, die cast metal, c. 1990s. $3-$15.

Roberts, Ride On, fire rescue van, pressed steel, c. 1950s. $50-$300.

Sabra Toy, fire chief's car, die cast metal, c. 1960s. $25-$125.

Road Tough, Emergency Team, fire engine, die cast metal, c. 1990s. $3-$15.

Saunders, fire chief's car, plastic, c. 1950s. $25-$175.

Saunders, Super Searchlight Fire Truck, plastic, c. 1950s. $35-$250.

Schaper, Heavy Haulers, stomper fire engine, die cast, c. 1980s. $3-$18.

Smith Miller, Model "L" Mack truck, hook and ladder, die cast metal, 35.75", c. 1950s. $175-$750.

Smith Miller, close-up of Model "L" Mack cab.

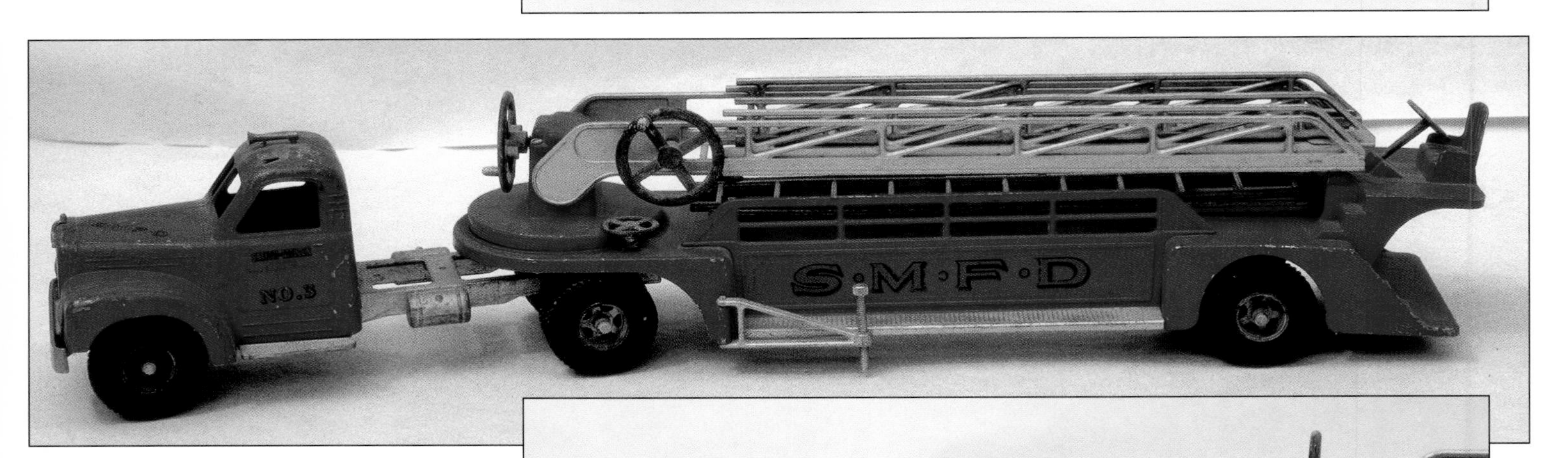


Smith Miller, Model "B" Mack truck, hook and ladder, cast metal, 35.75", c. 1954. $200-$850.

Smith Miller, close-up of Model "B" Mack cab.

Smith Miller, MIC aerial ladder, cast metal, 26", c. 1950s. $175-$700.

Solido, super pumper, die cast metal, c. 1980s. $10-$35.

Solido, antique Ford pickup, die cast metal, c. 1990s. $10-$50.

Solido, antique Ford tanker, die cast metal, c. 1990s. $10-$50.

Speedy Boat, fire boat, plastic, c. 1970s. $3-$20.

Structo, pumper, pressed steel, 22.5", c. 1950. $75-$200.

Structo, hook and ladder, pressed steel, 29.25", c. 1951-52. $75-$200.

Springbok, fire station puzzle, c. 1980s. $3-$8.

Structo, pumper, pressed steel, 13", c. 1960s. $50-$100.

Structo, fire rescue truck, pressed steel, c. 1960s (Note: Reissued in 1990s). $35-$125.

Structo, emergency rescue squad, pressed steel, c. 1960s (Note: Reissued in 1990s). $35-$125.

Structo, aerial hook and ladder, pressed steel, c. 1960s. $25-$150.

Sun Rubber Company, Mickey Mouse fire engine, rubber, 6.75", c. 1950s. $35-$125 shown.

TS, municipal fire department truck, cast metal, 9", c. 1960s. $25-$75.

Texaco Gas Station, fire chief speaker hat, plastic, c. 1960s. $25-$175.

Texaco Gas Station pump advertisement for fire chief hat (rare). $175-$500.

Texaco Gas Station giveaway, felt, 13.25" long, c. 1950s. $5-$25.

Tomy-Long Tomica, American LaFrance hook and ladder, die cast metal, c. 1980s. $10-$35.

Tomy-Tomica, American LaFrance ladder chief, die cast, c. 1970s. $5-$45.

Tomy Pocket Cars, emergency van, die cast, c. 1970s. $3-$20.

Tomy Pocket Cars, American LaFrance ladder, die cast, c. 1990s. $3-$20.

Tomy Pocket Cars, American LaFrance fire engine, die cast, c. 1970s. $3-$20.

Tonka, aerial ladder, No. 700-b, pressed steel, c. 1956. $60-$250.

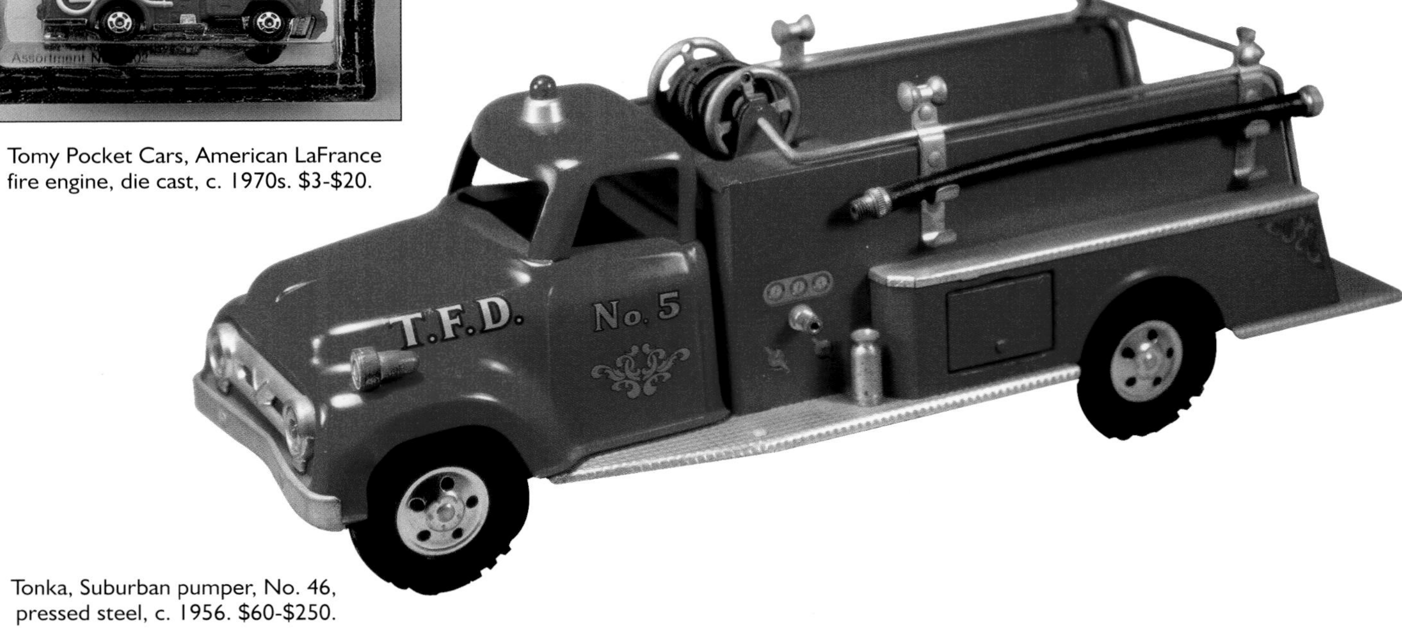


Tonka, Suburban pumper, No. 46, pressed steel, c. 1956. $60-$250.

Tonka, Suburban pumper, No. 46, pressed steel, c. 1957. $60-$250.

Tonka, hydraulic aerial ladder truck, No. 48, pressed steel, c. 1957. $60-$250.

Tonka, tanker from set 3212, rare, pressed steel, c. 1958. $150-$600.

Tonka, rescue van from set 3212, rare, pressed steel, c. 1958. $95-$350.

Tonka, Suburban pumper from set 3212, rare, pressed steel, c. 1958. $100-$350.

Tonka, Suburban pumper, No. 926, pressed steel, c. 1960. $75-$175.

Tonka, Suburban pumper, pressed steel, c. 1960s. $35-$125.

Tonka, fire chief jeep, pressed steel, c. 1960s. $25-$135.

Tonka, Jeep pumper, pressed steel, c. 1960s. $25-$135.

Tonka, Jeep pumper, No. 425, pressed steel, c. 1960s. $50-$275.

Tonka, pumper, No. 595, pressed steel, 6", c. 1975. $5-$25.

Tonka (front), pumper, No. 595, 6"; (rear) aerial ladder, pressed steel, 11", c. 1975. F: $5-$25; R: $5-$25.

Tonka, rescue van, pressed steel, 6", c. 1970s. $5-$20.

Tonka, fire chief van, pressed steel, 6", c. 1970s. $5-$20.

Tonka, aerial ladder, No. 1962, pressed steel, c. 1970s. $15-$80.

Tonka, snorkel, pressed steel, c. 1960s-70s. $25-$200.

Tonka, snorkel, pressed steel, c. 1970s. $25-$175.

Tonka, aerial ladder, c. 1960s-70s. $25-$250.

Tonka, fire pumper, pressed steel, c. 1970s. $15-$75.

Tonka, firefighter fire engine, pressed steel, c. 1960s. $15-$35.

Tonka, fire truck, pressed steel, c. 1980s. $5-$35.

Tonka, heavy duty pumper, No. 1360, pressed steel, c. 1980s. $5-$30.

Tonka, heavy duty pumper, No. 1360 (variation), pressed steel, c. 1980s. $5-$30.

Tonka, Mighty Hook & Ladder, No. 3977, pressed steel, c. 1980s. $15-$85.

Tonka, aerial ladder, No. 1875, pressed steel, c. 1980s. $15-$75.

Tonka, aerial ladder team, No. 1935, pressed steel, c. 1980s. $15-$75.

Tonka, rescue vehicle, No. 1386, pressed steel, c. 1980s. $15-$40.

Tonka, aerial ladder, No. 676, pressed steel, c. 1980s. $8-$25.

Tonka, pumper team, No. 2045, pressed steel, c. 1980s. $10-$45.

Tonka, Play People fireman, plastic, c. 1980s. $5-$15.

Tonka, Scramblers pumper, No. 1100, pressed steel, c. 1980s. $4-$15.

Tonka, Scramblers ladder, pressed steel, c. 1970s. $4-$12.

Tonka, Scramblers pumper, pressed steel, c. 1970s. $5-$18.

Tonka, Scramblers fire chief's car, No. 1100, pressed steel, c. 1980s. $8-$15.

Tonka, Scramblers pumper, No. 1100, pressed steel, c. 1980s. $5-$15.

Tonka, fire department set, No. 1009, pressed steel, c. 1980s. $35.

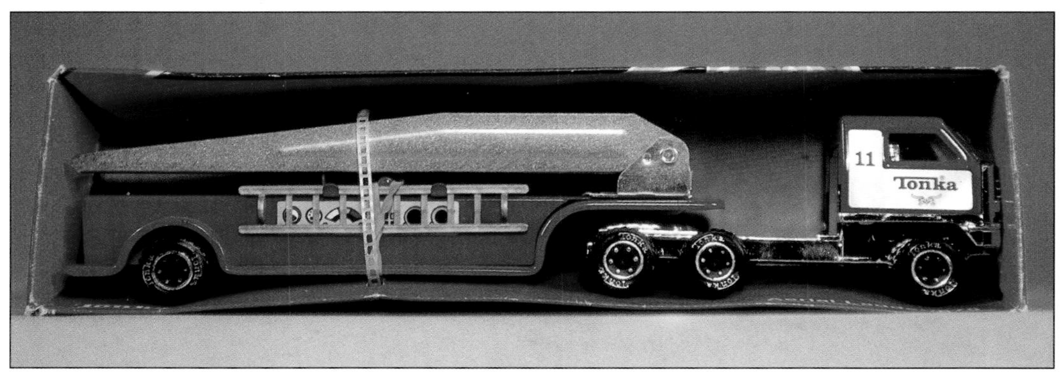


Tonka, aerial ladder, pressed steel, c. 1980s. $4-$12.

Tonka, Tough Treads pumper, c. 1990s. $2-$10.

Tonka, fire house play set, No. 5047, pressed steel and plastic, c. 1980s. $35.

Tonka, My 1st Tonka play set, fire house, plastic, c. 1990s. $15.

Tonka, GoBots Mighty Robot pumper, die cast, c. 1980s. $8-$35.

Tootsietoy, pumper, cast metal, 3", c. 1940s. $25-$60.

Tootsietoy, Ahrens-Fox pumper (two varieties), cast metal, both 3", c. 1940s. $25-$75 each.

Tootsietoy, pumper, cast metal, 3", c. 1940s. $25-$60.

Tootsietoy, ladder truck (two varieties), cast metal, 3", c. 1940s. $25-$60.

Tootsietoy, fire chief's car, cast metal, 4", c. 1950s. $5-$35.

Tootsietoy, fire department set, cast metal, c. 1950s. $200-$375.

Tootsietoy, "L" model Mack hook and ladder, cast metal, c. 1950s. $35-$100.

Tootsietoy, International Truck Company hook and ladder, cast metal, 8", c. 1950s. $50-$100.

Tootsietoy, "L" model Mack pumper, cast metal, c. 1950s. $25-$100.

Tootsietoy (front), Insurance Patrol, No. 337, cast metal, 3", c. 1940s; (rear) hose wagon, No. 238, cast metal, 3", c. 1940s. F: $15-$35; R: $15-$35.

Tootsietoy, fire engine, cast metal, c. 1978. $5-$35.

Tootsietoy, Toughs fire engine, die cast metal, c. 1970. $6-$20.

Tootsietoy, Special Missions emergency truck, die cast metal, c. 1970s. $7-$45.

Tootsietoy, Tiny Toughs, fire engine, die cast metal, c. 1971. $10-$35.

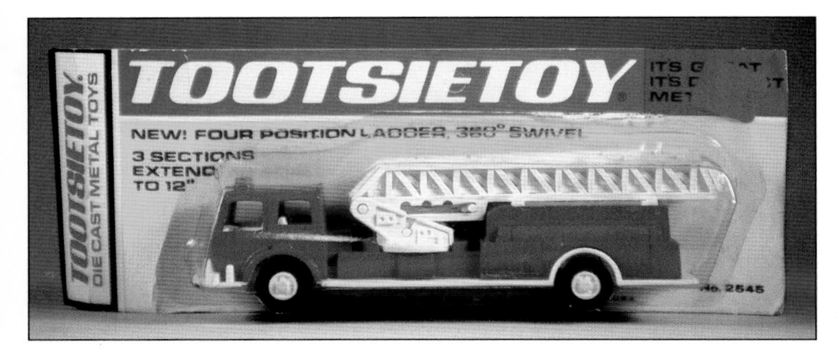


Tootsietoy, ladder truck, die cast metal, c. 1970s. $10-$35.

Tootsietoy, Toughs, utility unit, die cast, c. 1970s. $6-$25.

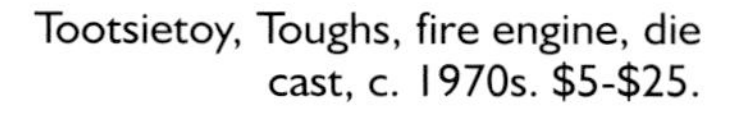

Tootsietoy, Toughs, fire engine, die cast, c. 1970s. $5-$25.

Tootsietoy, Toughs fire engine, die cast metal, c. 1970. $5-$25.

Tootsietoy, firefighters play set, die cast, c. 1970s. $35.

Tootsietoy, Toughs, fire engine, die cast, c. 1970s. $5-$20.

Tootsietoy, firefighters play set, die cast, c. 1970s. $30.

Totem, firefighter candy dispenser, plastic, c. 1960s. $20-$95.

Toyco, remote control fire boat, plastic, c. 1980s. $25-$95.

Toyland (rear), fire station, pressed steel, c. 1950s; Hubley (front), fire chief's car, No. 305, 6", plastic, c. 1940s. R: $50-$175; F: $15-$25.

Toy State, battery operated Mack fire engine, plastic, c. 1970s. $25-$45.

Traffic Stoppers, Engine House No. 4, die cast, c. 1990s. $20.

Traffic Stoppers, Engine House No. 5, die cast, c. 1990s. $25.

Transistor Robots, rescue squad, die cast, c. 1980s. $5-$15.

Tuco Puzzle, Mack fire engine, c. 1950s. $15-$35.

Tyco, U. S. 1 electric trucking fire station, plastic, c. 1970s. $15-$75.

Unknown manufacturer, Walt Disney Pop-A-Part fire engine toy game, cardboard and plastic, c. 1950s. $20-$150.

Unknown manufacturer, Mickey Mouse fire engine, plastic, c. 1990s. $5-$10.

Unknown manufacturer, Mickey Mouse fire engine, 7", available only at Disney theme parks, c. 1990s. $10-$25.

Unknown manufacturer, Mickey Mouse fire engine, tin, park item, c. 1980s. $15-$45.

Unknown manufacturer, fire department ambulance, tin, 13.5", c. 1950s. $45-$125.

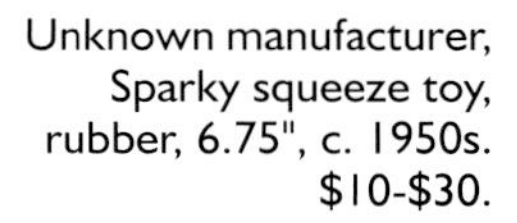

Unknown manufacturer, Sparky squeeze toy, rubber, 6.75", c. 1950s. $10-$30.

Unknown manufacturer, fire station bank, tin, 5", c. 1950s. $25-$75.

Unknown manufacturer, Santa on fire engine, plastic, 6.5", c. 1950s. $35.

Unknown manufacturer, fire engine, cast metal, c. 1950s. $15-$65.

Unknown manufacturer, alarm box bank, cast iron, 4.25" high, c. 1960s. $5-$50.

Unknown manufacturer, fire chief's car, friction, cast metal, 6", c. 1940s. $20-$75.

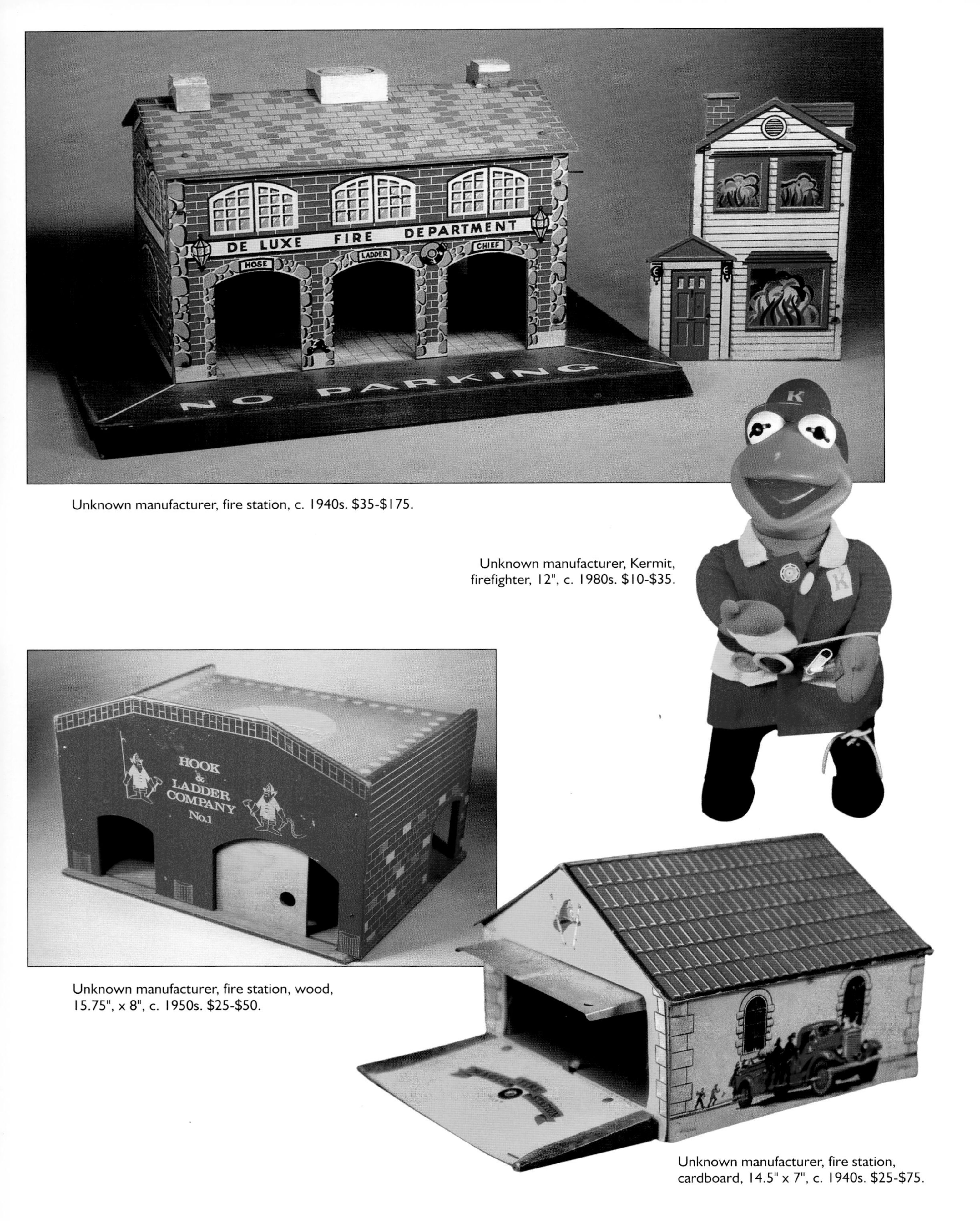


Unknown manufacturer, fire station, c. 1940s. $35-$175.

Unknown manufacturer, Kermit, firefighter, 12", c. 1980s. $10-$35.

Unknown manufacturer, fire station, wood, 15.75", x 8", c. 1950s. $25-$50.

Unknown manufacturer, fire station, cardboard, 14.5" x 7", c. 1940s. $25-$75.

Unknown manufacturer, Sesame Street fire department, cardboard, rubber, c. 1980s. $10-$35.

Unknown manufacturer, firefighter set, plastic, c. 1950s; firehouse 6", truck 4". $5-$20.

Unknown manufacturer, Old Smokey, cast metal, c. 1950s. $5-$15.

Unknown manufacturer, Snoopy firefighter bank, plastic, c. 1960s. $10-$45.

Unknown manufacturer, wooden fire engine and chief's car, both 4.75", c. 1980s. $5-$15 each.

Unknown manufacturer, possibly Hallmark, Rovesville Firehouse, c. 1980s. $20.

Unknown manufacturer, snap-on fire rescue team, plastic, c. 1970s. $15.

Unknown manufacturer, puzzle, firefighters, c. 1970s. $10-$25.

Unknown manufacturer, action outfits, crash truck outfit, c. 1970s. $5-$15.

Unknown manufacturer, Huckleberry Hound, firefighter squeak toy, rubber, 8.5", c. 1970s. $25-$50.

Unknown manufacturer, Howdy Doody squeeze toy, rubber, 5.75", c. 1950s. $75.

Unknown manufacturer, Snoopy pencil case, fire engine, plastic, c. 1960s. $5-$10.

United Feature Syndicate, large size Snoopy in fireman outfit, c. 1970s. $25-$75.

Walt Disney Company, Disney fire boat, plastic, c. 1990s. $8-$35.

United Feature Syndicate, Snoopy's wardrobe, firefighter, large and small size, c. 1970s. $10-$25 each.

Waterline Boat, fire boat, plastic, c. 1970s. $10-$35.

Whitman, Guy Smiley firefighter puzzle, c. 1980s. $5-$15.

Whitman, Goofy firefighter puzzle, c. 1970s. $10-$35.

Whitman, Lassie puzzle, c. 1972. $25-$50.

Wyandotte, hook and ladder, tin, 12", c. 1940s/50s. $25-$65.

Wyandotte, Toytown fire house, tin, 6.5" x 5", c. 1950s. $35-$125.

Zee Toys, The Hustlers fire engine, plastic, c. 1970s. $10-$35.

Zee Toys, Mini Macks snorkel, die cast, c. 1970s. $3-$15.

Zee Toys, Mini Macks, hook and ladder, die cast, c. 1970s. $3-$15

Chapter Two

# Smokey the Bear Collectibles

A symbol of the U.S. Fire Service, Smokey the Bear initially came about through a creative advertising campaign. In 1944, the Wartime Advertising Council provided specifications for a public service poster that would promote the safe use of our National Forests. A brown bear, combination forest ranger, was the result. He would be called "Smokey" and would teach about good citizenship in the woods and forests. His friendly and fatherly, good-natured advice became well known via the slogan: "only YOU can prevent forest fires."

A tiny bear cub that was rescued from a fire and nursed to health by a forest ranger became a real-life Smokey the Bear in 1950. In adulthood, Smokey took up residence at the zoo in Washington, D.C. and was so popular that, at one time, he received approximately 13,000 letters per month. Today, the zoo houses a successor to the original Smokey so that Americans will be able to continue to visit a real live Smokey the Bear.

Collectibles in the Smokey genre are very popular. Banks and all manner of toys are continuously manufactured.

Dakin, Smokey Bear giveaway, 8", plastic, c. 1976. $18-$35.

Gabriel, Smokey Bear Jeep, die cast, c. 1970s. $20-$100.

Lakeside, Smokey Bear, super flex, c. 1970s. $20-$60.

Ideal, Smokey Bear plush doll (missing hat), rubber face, c. 1950s. $25-$200.

Larami, Smokey Bear camping set, c. 1978. $65.

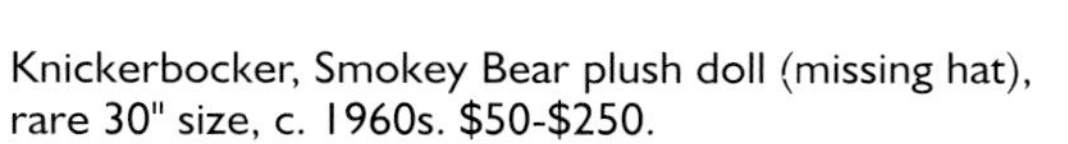

Knickerbocker, Smokey Bear plush doll (missing hat), rare 30" size, c. 1960s. $50-$250.

Milton Bradley, Smokey Bear game, c. 1960s. $20-$75.

Unknown manufacturer, Smokey Bear plastic banks, c. 1960s. L: $45-$100; R: $40-$90.

Unknown manufacturer, set of three miniature china Smokey Bear figures, c. 1960s. $35-$75.

Unknown manufacturer, Smokey Bear milk mug, c. 1960s. $15-$35.

Unknown manufacturer, Smokey Bear pajama bag, c. 1960s. $25-$50.

Chapter Three

# Firefighting Models

Plastic toy model kits, a toy/hobby/skill item that found its apex around 1955, were traditionally marketed to boys in the form of toy fire engines, cars, trucks, airplanes, etc. Companies such as Aurora, Revell, Monogram, and Lindberg challenged young men into crafting images of their favorite vehicles. After the middle of the 1950s, toy model kits became highly diversified as to model topic. Monster figures, popular character figures, and figural scenes became all the rage. Today, some share of the model kit market remains with the ever-popular wheeled vehicle.

AMT, fire chief's car, c. 1970s. $5-$40.

AMT, fire rescue van, c. 1970s. $5-$40.

AMT, American LaFrance ladder chief, c. 1970s. $25-$75.

AMT, American LaFrance custom pumper, c. 1970s. $25-$100.

AMT, American LaFrance aero chief, c. 1970s. $25-$100.

AMT, reissue American LaFrance ladder chief, c. 1980s. $10-$25.

AMT, vintage Ford chief's car. $10-$25.

Ashton, Ahrens Fox quint metal, c. 1970s. $50-$175.

AMT, Toyota fire chief's car, 4x4. $10-$25.

Ashton, Ahrens Fox piston pumper, metal, c. 1970s. $50-$175.

Ashton, Ahrens Fox pumper, metal, c. 1970s. $50-$125.

Aurora, Hot Surfer custom pumper, c. 1960s. $50-$200.

Aurora, American LaFrance 900 pumper, c. 1960s. $50-$150.

Big Wheels, American LaFrance pumper. $5-$15.

Conrad, aerial ladder truck, metal, 17", c. 1980s. $50-$175.

Conrad, Emergency One, 3 axle aerial ladder, metal, c. 1980s. $50-$175.

Conrad, Emergency One airport crash truck, metal, c. 1980s. $50-$195.

Conrad, LTI ladder truck, metal, c.1980s. $35-$190.

CraftMaster, Red Rover fire engine, wood, c. 1960s. $35-$150.

Gabriel, 1914 Model T chemical hose truck, metal, c. 1970s. $35-$100.

Enter, reissue fire boat firefighter, c. 1990s. $10-$35.

ITC, fire pumper, c. 1960s. $5-$15.

Jordan, HO highway miniatures, c. 1970s. $10-$25 each.

Jo-Han, fire rescue ambulance, c. 1980s. $10-$35.

Lindberg, American LaFrance series 900 pumper, c. 1960s. $10-$45.

Lindberg, American LaFrance series 900 pumper, c. late 1960s. $10-$40.

Lindberg, American LaFrance series 900 pumper, white, c. late 1960s. $10-$75.

Lindberg, fire boat, c. 1960s. $10-$35.

Lindberg, FDNY Capt. T. Ahern fire boat, c. 1960s. $10-$50.

Lindberg, mini American LaFrance pumper, c. 1960s. $10-$25.

Marlin, American firefighters, c. 1952. $10-$35.

Monogram, Mack fire engine, c. 1980s. $10-$20.

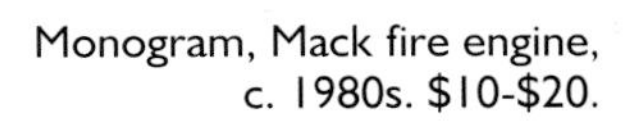
Monogram, Mack fire engine, c. 1980s. $10-$20.

Monogram, fire chief's car issued for *Backdraft* movie, c. 1990s. $10-$25.

Monogram, Mack pumper, reissued for *Backdraft* movie, c. 1990s. $10-$25.

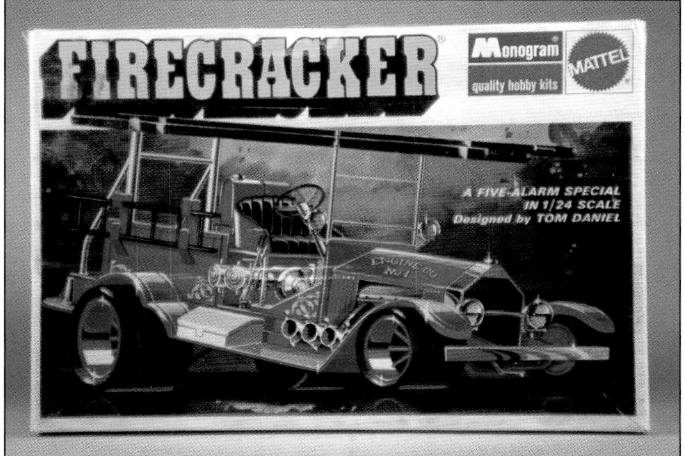


Monogram, Firecracker custom fire engine, c. 1970s. $20-$60.

Monogram, Fire Iron custom fire engine, c. 1970s. $20-$60.

Monogram, fire helicopter, c. 1970s. $10-$25.

MPC, Chevy pickup, c. 1970s. $10-$25.

MPC, 1940 Ford fire chief's car, c. 1980s. $10-$25.

MPC, Christie front drive steam fire engine, c. 1970s. $25-$60.

MPC, custom fire truck, c. 1970s. $20-$35.

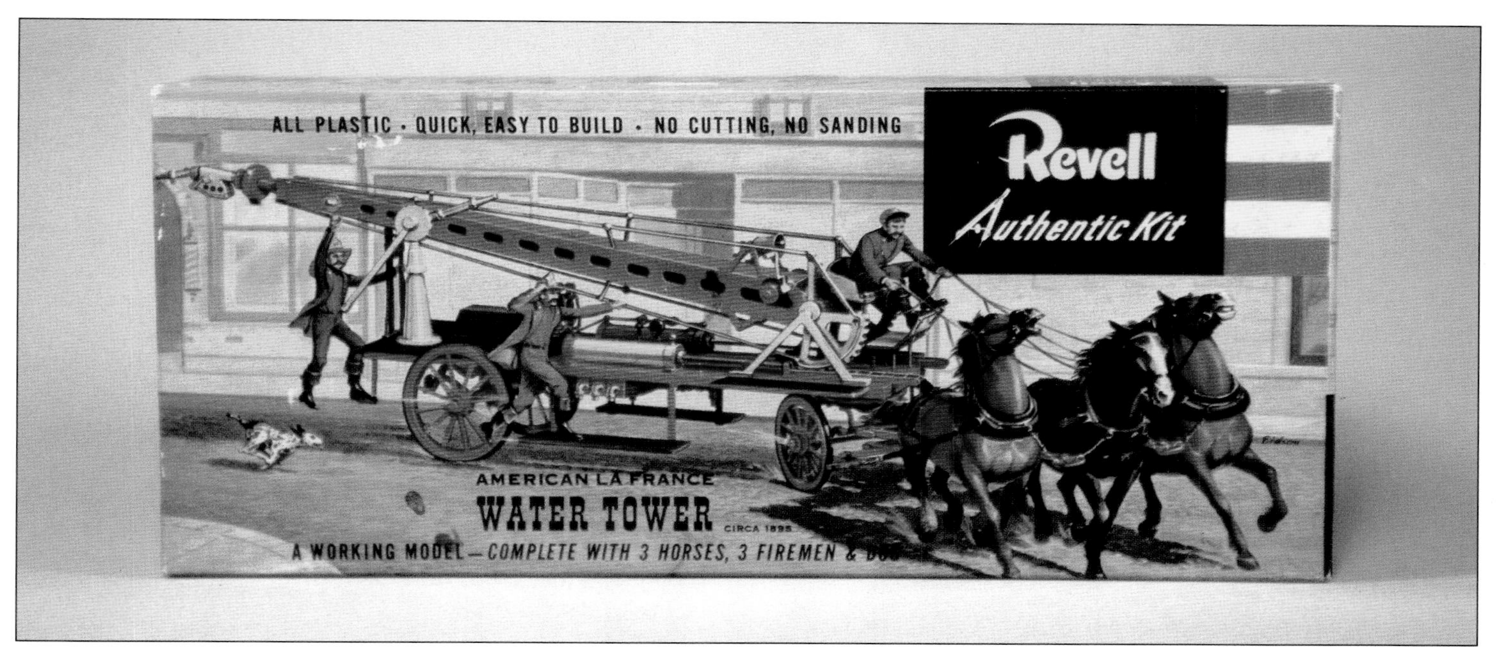


Revell, water tower, c. 1960s. $50-$125.

Revell, hook and ladder (rare in this box). $50-$200.

Revell, American firefighters steamer, c. 1960s. $15-$50.

Skyline, fire house, c. 1940s. $20-$50.

Unknown manufacturer, Cinder Bug custom fire engine, c. 1960s. $20-$35.

Tonka, diorama alarm call, c. 1960s. $25-$90.

Unknown manufacturer, Firebug custom fire engine, c. 1960s. $20-$35.

Unknown manufacturer, American LaFrance pumper, metal, c. 1970s. $45-$90.

Unknown manufacturer, early American fire engine, c. 1960s. $5-$20.

Wilo Line, fire truck kits, rare, c. 1950s. $50-$200 each.

Chapter Four

# TV Show Collectibles

Although there have been many TV shows featuring a firefighting theme format, we will focus on two, namely, *Code Red* and *Emergency*. These shows were fairly significant productions and through these we are able to illustrate the TV related firefighting collectible toy.

***Code Red*** was a short-lived, though educational and dramatic TV series focusing on a fictional Los Angeles family of firefighters and arson investigators. The ninety-minute pilot aired on September 20, 1981, and was followed by eighteen one-hour episodes concluding in July of 1982. The show's unfortunate juxtaposition opposite *60 Minutes* may have contributed to a lack of viewer interest and to its ultimate demise. Show collectibles include: Matchbox scale models; a *Code Red* hook and ladder by Processed Plastic Company; kits for fire fighters, fire investigators, and paramedics; Imperial Toy Company's plastic fire fighters; and model fire vehicles by Revel.

Buddy L, Code Red rescue set, pressed steel. $25-$75.

Matchbox, LAFD fire boat, die cast. $5-$18.

Buddy L, Code Red hook and ladder, pressed steel. $10-$25.

Matchbox, ambulance, die cast. $5-$18.

Matchbox, LAFD chopper, die cast. $5-$18.

Matchbox, LAFD snorkel, die cast. $5-$18.

Matchbox, LAFD chief's car, die cast. $5-$18.

Matchbox, LAFD engine, die cast. $5-$18.

Placo Toys, Battalion Chief Playset, plastic. $15-$75.

Processed Plastics, firefighter figure set, plastic. $2 each or $75 package.

Processed Plastics, LAFD snorkel with six figures, plastic. $10-$35.

Revell, model LAFD chief's car, plastic. $10-$45.

Revell, model LAFD rescue van, plastic. $10-$40.

Revell, model LAFD rescue chopper, plastic. $10-$40

***Emergency***, which aired from 1971 to 1977, was a TV program that dramatized and popularized a new and radical use of rescue personnel in the U.S.: that of the firefighter-paramedic.

The use of a first responder to medical emergencies was a fairly new concept back in 1971, when filming began for the movie pilot of *Emergency*. In fact, Governor Ronald Reagan had only just signed the Wedworth Townsend Act in 1970, which allowed first responders to act without nurses.

Exciting and educational, *Emergency* did a lot to educate the public on medical safety needs and issues. An extremely popular show, *Emergency* was aired in forty-one countries by 1976 and was considered one of the top ten shows of that era. Collectibles abound; one may find *Emergency* board games, lunch boxes, View Masters, puzzles, comics, helmets, record albums, walkie talkies, medical kits, and more.

Aladdin lunch box, metal.
$20-$250 mint.

Aladdin lunch box (opposite side) and thermos, plastic. $15-$35 thermos only.

Collegeville, Emergency fireman Halloween costume. $25-$75.

Collegeville, original art for costume advertisements. $175.

Dinky, Emergency paramedic truck, No. 77, die cast. $25-$300.

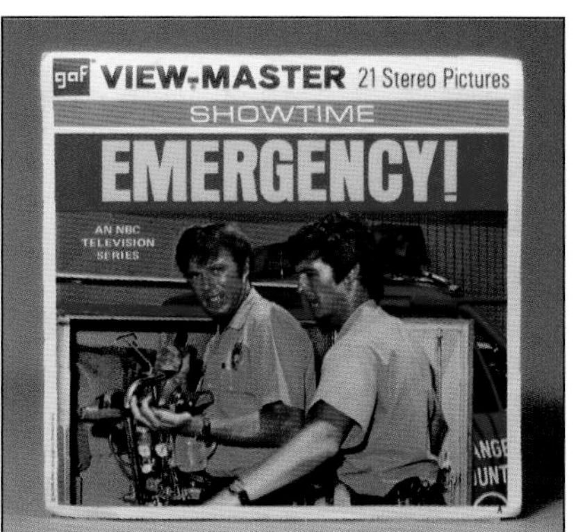

G.A.F., Emergency view-master. $15-$35.

Fleetwood Toys, Emergency rescue squad with fire house, plastic. $75-$125.

LJN, Emergency play set. $35-$200.

LJN, Roy and John action figures. $15-$100.

LJN, Roy and John action figures, individual packs. $15-$75.

LJN, Roy action figure (package variation). $75.

LJN, Emergency rescue truck, plastic, for action figures (rare). $50-$300.

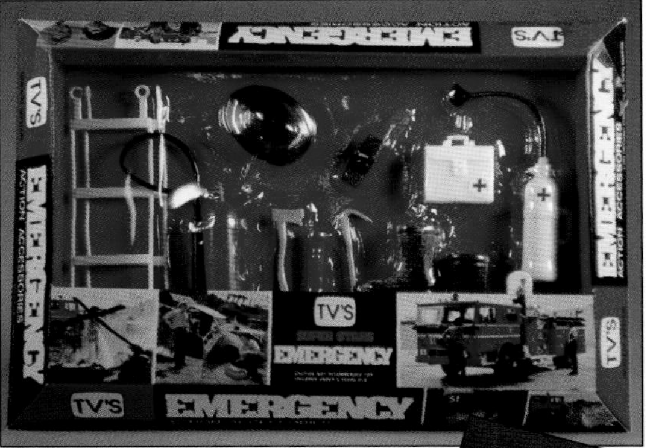


LJN, accessories for action figures. $75.

LJN, Emergency fire engine, die cast, $10-$40.

LJN, Emergency rescue truck, die cast. $10-$40.

Milton Bradley, Emergency game. $25-$60.

PLACO Toys, Emergency red plastic helmet. $10-$20.

PLACO Toys, Emergency black plastic helmet, rare. $20-$45.

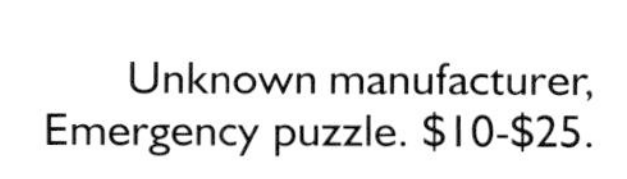

Unknown manufacturer, Emergency puzzle. $10-$25.

Unknown manufacturer, set of Emergency puzzles. $10-$35 each.

# Bibliography


Bruegman, Bill. *Aurora History and Price Guide*. Ohio: Cap'n Penny Productions, Inc., 1994.

Cain, Dana. *Saturday Morning TV Collectibles—60's, 70's, 80's*. Wisconsin: Krause Publications, 2000.

DeSalle, Don & Barb. *The DeSalle Collection of Smith-Miller & Doepke Trucks*. Indiana: L-W Book Sales, 1997.

Force, Dr. Edward. *Corgi Toys*. Pennsylvania: Schiffer Publishing Ltd., 1984.

Force, Dr. Edward. *Dinky Toys*. Pennsylvania: Schiffer Publishing Ltd., 1988.

Fox, Bruce R., and John J. Murray. *Fisher- Price - Historical, Rarity and Value Guide, 1931-Present.* Wisconsin: Krause Publications, 2002.

Hake, Ted. *Guide to Advertising Collectibles.* Pennsylvania: Wallace-Holmestead Book Company, 1992.

Hake, Ted. *Hake's Guide to TV Collectibles*. Pennsylvania: Wallace-Holmestead Book Company, 1990.

Hake, Ted. *Hake's Price Guide to Character Toys*. New York: Gemstone Publishing, Inc., 2002.

Hansen, Charles V. *The History of American Firefighting Toys*. Maryland: Greenberg Publishing Company, Inc., 1990.

Korbeck, Sharon, and Stearns, Clan., Editors. *Toys and Prices 2003*. Wisconsin: Krause Publications, 2002.

O'Brien, Karen, Editor. *Toys & Prices 2004*. Wisconsin: Krause Publications, 2004.

O'Brien, Richard. *Collecting Toy Cars &Trucks—Identification and Value Guide*. Wisconsin: Krause Publications, 1997.

Rich, Mark. *Toys A to Z: A Guide and Dictionary for Collectors, Antique Dealers and Enthusiasts.* Wisconsin: Krause Publications, 2001.

Smith, Michelle L. *Marx Toys Sampler—A History and Price Guide*. Wisconsin: Krause Publications, 2000.

Stephan, Elizabeth A., Editor. *O'Brien's Collecting Toy Cars & Trucks. Identification & Value Guide*. Wisconsin: Krause Publications, 2000.

Turpen, Carol. *Baby Boomer Toys and Collectibles*. Pennsylvania: Schiffer Publishing Ltd., 2002.

Yokley, Richard C. *T.V. Firefighters*. California: Black Forest Press, 2002.